Ondřej Filipec

REACH Beyond Borders

Europeanization Towards Global Regulation

Springer

Ondřej Filipec
Faculty of Social Sciences
University of Ss. Cyril and Methodius
Trnava, Slovakia

ISBN 978-3-319-54153-2 ISBN 978-3-319-54154-9 (eBook)
DOI 10.1007/978-3-319-54154-9

Library of Congress Control Number: 2017945760

Printed on acid-free paper

This Springer imprint is published by Springer Nature
The registered company is Springer International Publishing AG
The registered company address is: Gewerbestrasse 11, 6330 Cham, Switzerland

Dedicated to All Victims of Hazardous Chemicals

Contents

Abbreviations

BPR	Biocidal Product Regulation
CCA	Chemicals Control Act (Korea)
CEPA	Canadian Environmental Protection Act
ChemO	Chemicals Ordinance on Protection Against Dangerous Substances and Preparations (Switzerland)
CICR	Regulation on the Inventory and Control of chemicals (Turkey)
CLP	Classification, Labelling and Packaging
CSIA	Chemical Safety Improvement Act (USA)
EHSNR	Environmentally Hazardous Substances Notification and Registration (Malaysia)
EINECS	European Inventory of Existing Commercial Chemical Substances
EPA	Environmental Protection Agency (USA)
GOSTs	Product-Related-specific Mandatory Standards (Russia)
HSNO	Hazardous Substances and New Organisms Act (New Zealand)
NICNAS	National Industrial Chemicals Notification and Assessment Scheme (Australia)
PBT	Persistent, Bioaccumulative and Toxic substances
PIC	Prior Informed Consent
POPs	Persistent Organic Pollutants
REACH	Regulation (EC) No 1907/2006 of the European Parliament and of the Council of 18 December 2006 concerning the Registration, Evaluation, Authorisation and Restriction of Chemicals
SCA	Safe Chemicals Act (USA)
TCCA	Korean Toxic Chemicals Control Act.
TCSCA	Toxic Chemicals Substances Control Act (Taiwan)
TSCA	Toxic Substances Control Act (USA)
TSHNWC	Toxic Substances and Hazardous and Nuclear Waste Control Act (Philippines)
vPvB	Very Persistent and Very Bioaccumulative

Introduction

The story of REACH[1] was very dramatic from the early beginning. After the European Union finished formally the single market by adopting the Maastricht Treaty in 1993, it was self-evident that the common market requires also stronger common rules. This was also the case of chemicals where national policies were only partly harmonized through a series of directives which were considered insufficient already at the time they were enacted. There was a significant lack of information about substances placed on the market. Old systems of chemical management were limited in data gathering, and there was only little information about known effects of substances on human health and environment. The burden of proof has been placed on consumers who, in the case of harm, only hardly defend their rights during trials with giant corporations and chemical producers. At the same time, the volume of trade between EU member states was increasing, and new products were placed on the EU market without sufficient knowledge about their possible long--term effects. Rigid, obsolete and insufficient directives resulted in insufficient regulation of chemicals which was strongly criticised by NGOs, politicians and sometimes even by the chemical industry itself. The need for substantial reform was recognised by all actors including the chemical industry

The negotiations about new regulation took almost a decade. It was really a very long way leading through opposing and sometimes acrimonious interests of thousands of actors. The lobbing of chemical industry and environmental NGOs to affect institutions and national interest was permanent. The intensity and cruel practices used resulted in REACH being referred to as the most lobbied regulation

[1] In full reference Regulation (EC) No. 1907/2006 of the European Parliament and of the Council on 18 December 2006 concerning the Registration, Evaluation, Authorisation and Restriction of Chemicals (REACH) establishing European Chemicals Agency, amending Directive 1999/45/EC and repealing Council Regulation (EEC) No. 793/93 and Commission Regulation (EC) No. 1488/94 as well as Council Directive 76/769/EEC and Commission Directives 91/155/EEC, 93/67/EEC, 93/105/EC and 2000/21/EC (further abbreviated simply as REACH).

in EU history.[2] The war between actors over new chemical legislation lasted almost a decade and in some aspects is still ongoing. REACH entered into the force in June 2007. Since then EU became an entity with the most comprehensive and ambitious chemical regulation in the world, establishing a new paradigm within environmental policy.

However, even after REACH entered into force in 2007, its implementation did not stop to cause concern among chemical companies, especially small and medium size enterprises. Moreover, REACH was only beginning. The EU regulatory landscape has been enriched by Classification, Labelling and Packaging (CLP) legislation in 2009 and by Biocidal Product Regulation (BPR) and the Prior Informed Consent Regulation (PIC) by 2012. REACH regulation is still being developed as the review process left some space for regulation of substances produced in low volumes, endocrine disruptors or nanomaterials. The EU regulatory landscape regarding chemicals changed in a revolutionary way and became the source of inspiration for other countries in the world.

Research Aims

The main aim of this book is to explore the EU influence in the area of chemical regulation on the other parts of the world and assess how much other countries are inspired or influenced by it. This will be analysed through the concept of Europeanization which is suitable for such analysis with great potential for application and deriving new interesting implications. The book will not only explore the extent of how much national policies are Europeanized but will also search for possible political and economic causes, such as the size of chemical industry, the trade between EU and the country, export orientation or other influences. This is very important in order to reveal the nature of EU influence beyond its borders and verify the extent in which REACH influenced global perception on chemical management.

For this reason, the main hypothesis of the book is that: "The level of economic interdependence between the EU and selected non-EU states in the field of chemicals positively correlates with the Europeanization level of national chemical regulation". In other words, the more non-EU country is exporting to the EU, the more Europeanized (REACH-like) regulation there will be. As pointed by Steven Levitsky and Lucan Way (2006), the density of ties and cross-border flows between a particular country and the EU may play an important role in determining the level of influence of the EU (Levitsky and Way 2006: 383).

In the context of Europeanization, chemical regulation is seen as a dependent variable and the level of economic interdependence as independent variable. Nevertheless, other variables may intervene in this relationship (e.g. similarity of

[2] For further details see for example Contiero M (2006) Toxic lobby: how the chemicals industry is trying to kill REACH. Greenpeace International) or Selin H (2007) Coalition politics and chemical management in a regulatory ambitious Europe. Glob Environ Polit 7(3):63–93.

domestic structures, geography, structure of chemical industry etc.). For this reason, different explanatory alternatives are also discussed. In order to reveal the level of Europeanization, chemical regulations of selected states will be explored and compared in the context of REACH. For this purpose, the structure of the book was adapted.

Structure of the Book and Methodological Aspects

The publication is divided into three parts. The first part is of theoretical nature and consists of three chapters. Chapter 1 explores the concept of Europeanization and its core elements which are further focused on in this book. Chapter 2 explores REACH as the source of Europeanization. If we want to know the answer of "how much" is something Europeanized, we have to explore the source of Europeanization in order to identify how possible Europeanization looks like. At the centre of this book is REACH; however CLP, BPR and PIC regulations are introduced as well. Aside from EU chemical regulation, there might be other sources of inspiration for countries outside EU which are presented in Chap. 3. These sources are present at an international level in the form of international treaties, conventions or other instruments and initiatives. For the sake of completeness, it is desirable to identify other sources of influence on the international level which may affect domestic chemical regulations outside Europe. This is partially important as it is necessary to distinguish the process of Europeanization and globalization.[3]

The second part of the book is more empiric in nature. It explores the chemical regulation in different countries of the world with significant chemical industry. Chapter 4 focuses on the EU candidate states (Turkey, Serbia, Montenegro, Albania and Macedonia) and potential candidates (Bosnia and Herzegovina, Moldova and Ukraine); Chap. 5 is dedicated solely to the USA where complex chemical reform has been considered for several years. Chapter 6 is dedicated to BRIC countries (Brazil, Russia, India and China). Chapter 7 deals with Japan, South Korea, Taiwan and Singapore and Chap. 8 with Malaysia, Indonesia, Thailand and the Philippines. Chapter 9 discusses other countries important for global chemical production, including Switzerland, Canada, Australia and New Zealand.

All chapters have similar structure. First, the status of chemical industry is explored in order to provide the reader a general context and deeper understanding of country's economic orientation in the field of chemicals. Second, the former chemical regulation is briefly described in its key aspects, and third, the new chemical legislation is explored. Because the legal landscape regarding chemicals is very heterogeneous and sometimes chaotic, the main focus is on the regulation of haz-

[3] Europeanization is often parallel with the process of globalization or modernization. Especially South-European states or Balkan states saw Europeanization as synonymum for modernization which legitimized domestic reforms or led to antagonistic reaction (see, e.g., Featherstone and Kazamias 2000, Anastakis 2005).

ardous chemicals and if possible the registration and evaluation of all industrial chemicals despite this approach among countries being still rare. The shift in the new chemical regulation is put into the context of EU REACH, and possible EU normative influences are identified.

The third part of the book consists only of one chapter (Chap. 10) which explores the practical and theoretical impacts of EU regulation on selected countries. In the first section, the impact of economic interdependence is considered as a key variable in the process of Europeanization. Other variables and models in context with chemical regulation are discussed in the second section.

The book focuses on the theoretical concept of Europeanization, which is presented in the first chapter. From the methodological point of view, it may be considered as an exploratory case study dealing with the regulation of hazardous chemicals. However, due to its complexity and territorial approach, one may suggest that it is not a single-case study dealing with chemical regulation but rather a set of smaller case studies (Part 2 of the book) exploring chemical regulation around the world and comparing it with the EU regulation. For this reason, each individual chapter dealing with national regulation has the same structure, enabling comparison with other legislations.

The State of Research

This book is mainly based on national chemical laws and treaties. However, in some aspects this rather specific topic has already been dealt with academically. Some publications were indispensable for the purpose of this book. Except "classical" theorists of Europeanization such as Robert Ladrech, Claudio Radaelli, Tanja Börzel and Thomas Risse, a very valuable contribution was made by Frank Schimmelfennig (2012) who examines Europeanization beyond the EU member states. This approach is crucial for this book, and because many authors focus on Europeanization vis-a--vis EU member states or the EU candidate states, the current book offers a unique opportunity to explore how Europeanization works beyond the borders of the EU. Europeanization is not a well-defined and clearly demarcated theoretical concept. For this reason, a very valuable contribution was presented by Timm Beichelt (2008) who presented three complex models for assessing Europeanization which were considered for the application, and Chap. 10 of this book builds on its aspects.

There are several pieces of literature regarding the empirical part of the book. In the case of the US regulation, the book focuses on the article written by John S. Applegate (1998) which compares TSCA and REACH. This article makes understanding of the US regulatory problems easier. Regulatory problems are well summarised also in another article written by Richard A. Denison (2009) who presented ten essential elements in TSCA reform. The reform process is well documented by Jerry H. Yen who wrote in 2013 a special report for the Congress in which reform proposal S. 1009 and S. 696 are compared and the reader may read provisions of all three acts (TSCA, CSIA and CSA) next to each other. The latest reform proposal –

The Frank R. Lautenberg Chemical Safety for the 21st Century Act – which entered into force in 2016 is not yet assessed at the academic level.

Despite the subject of chemical regulation being rare and resources scarce in the academic literature, there are several valuable contributions related to EU REACH influence. Very valuable contribution is presented by Yoshiko Naiki (2010) who analysed the reluctance of Japan and conditions unfavourable towards comprehensive REACH-like reform. Similarly important paper was presented by Tachwa Lee (2015) who analysed positive attitude of South Korea towards REACH-like regulation. Despite both articles having different approach towards analysis, various aspects of the articles are used and put into the context of Europeanization in this book.

An important source of information is the website of the UK Chemical Watch news server which is a rich resource for regulatory specialists dealing with chemicals. Another important source is ChemSafetyPro server, which was created by regulatory experts to meet important and detailed information about regulatory regimes regarding chemicals. Despite the fact that the author focused on primary sources of information (legal acts or treaties) in some countries, acts were written only in a native language. For this reason ChemSafetyPro server is a valuable source of unique information, as well as Chemical Inspection and Regulation Service (CIRS) website where regulatory experts publish overviews of chemical regulations.

Recommendation

Jacques Delors once reportedly said that "nobody can fall in love with the single market". The same can be said about chemical regulation and REACH which raises scepticism among companies rather than positive feelings. However, chemical regulation has many positive aspects and impacts on everyday lives of millions, and for this reason, it is worthy to explore its empirical and normative value. Publication is a valuable source of information for all scholars conducting their research in environmental policy, chemical regulation or chemical management as well as for regulatory experts in public and private institutions including NGOs, emerging chemical exporters, company lawyers or civil servants within the field of chemical management.

Comparing chemical regulations all over the world is a very demanding process requiring high level of multidisciplinary expertise. Moreover, chemical regulation is usually based on several pieces of law and implementing measures which may develop over time. For this reason, it is necessary to check validity of relevant regulatory parts within national regulatory regimes. In its relevant parts, this book refers to individual legal acts in order to avoid misinterpretation. Despite the many people helping me with this book, all mistakes are my own.

Prague
May 2, 2016

Ondřej Filipec

Part I
Theoretical Basis

Chapter 1
The Concept of Europeanization

This chapter explores key elements of the Europeanization[1] concept which is the main theoretical point of view used in this book. At first, several definitions of the concept will be introduced in order to explore different perceptions and understandings. Then emphasis is put on processes happening behind the label of Europeanization. Because these processes may evoke reactions of subjects to Europeanization or other actors influenced by Europeanization it is worth to explore possible responses to Europeanization at the domestic level. The last part of the chapter deals with the concept of Europeanization in the context of EU external influence as key mechanism used in this book.

1.1 Defining the Concept

The term Europeanization is relevant for almost three decades to describe various processes and phenomena dealing with the influence of Europe or in a more restrictive approach, the influence of the EU. As a number of authors used the term in their research the meanings became hazier and led to dozens of definitions. Definitions vary according to research area and variables highlighted in the concept of Europeanization. There is no shared concept of Europeanization as it may refer to different processes and attitudes.

For example, from a historical perspective Europeanization may refer to the export of European values and institutions outside the Europe as a result of colonial policies. Europeanization in this sense changed its substance by interaction with other cultures and development of identity (see Flockhart 2010). Others like John Borneman and Nick Flowler sees Europeanization as a consequence of the Cold War. While Western Europe was after the Second World War subject to

[1] In the scientific literature both terms „Europeization“ and „Europeanization“ are used. However, according to various scientific databases and Google the term “Europeanization” is much more common.

O. Filipec, *REACH Beyond Borders*, DOI 10.1007/978-3-319-54154-9_1

Americanization, Eastern Europe was subject of Sovietization. After the collapse of the Soviet Union both processes were overridden by Europeanization (Borneman and Fowler 1997). Borneman and Fowler similarly as Flockhart are not eliding Europeanization with the EU (Borneman and Fowler 1997: 492; Flockhart 2010: 790). Europeanization has a rather broad historical dimension connected to the identity. However, in this book Europeanization is used through the perspective of political science which is EU centric and where EU is the main actor in the process of Europeanization.

As pointed out by Johan P. Olsen (2002), in the context of the EU Europeanization is used in at least five different meanings. First, it refers to the changes in external boundaries and spread of EU system of governance. This is connected especially with the EU enlargement process. Second meaning refers to institutional development at the EU level with specific focus on centre-building institutions. Third meaning covers central penetration of national systems of governance and refers especially to adaptation to EU policies, values and norms (Regarding this point see Börzel 1999). Forth meaning of Europeanization refers to export of the norms of political organization and governance beyond European territory, especially due to international trade. Last meaning of Europeanization is connected to political unification project which combines in various degrees previous meanings under the more unified and stronger political entity (Olsen 2002: 923). This book reflects fourth meaning of Europeanization as the REACH may be considered system of governance or at least normative system which has international influence.

Because of different approaches towards Europeanization there are many definitions. For example Robert Ladrech in 1994 defined Europeanization as *"an incremental process reorienting the direction and shape of politics to the degree that EC political and economic dynamics become part of the organizational logic of national politics and policy-making"* (Ladrech 1994: 69). In some aspects this definition is simply comparable to the definition brought by Claudio Radaelli. Radaelli (2004) claims that *"Europeanization consists of processes of a) construction, b) diffusion and c) institutionalization of formal and informal rules, procedures, policy paradigms, styles, "ways of doing things" and shared beliefs and norms which are first defined and consolidated in the EU policy process and then incorporated in the logic of domestic (national and sub-national) discourse, political structures and public policies"* (Radaelli 2004: 5). In this book Europeanization is understood in a similar way as referred to by Simon Bulmer and Martin Burch (1998). In their view Europeanization could be seen in two dimensions. The Intrastate dimension refers to the impact of EU policies, rules, practices and values upon member state activities in respect of both the making and implementation of policy. It refers *"to the impact extent to which EC/EU requirements and policies have affected the determination of member states' policy agendas and goals. Secondly, it refers to the extent to which EU practices, operating procedures and administrative values have impinged on, and become embedded in, administrative practices of member states"* (Bulmer and Burch 1998: 602).

Despite the above mentioned definition being EU centric and referring to EU member states, which is influenced by the impact of EU policies, rules, values and practices; it could be applied to non-EU member states as well if we substitute EU member states simply by "states". This substitution, however, will of course have implications in the terms of direct link between EU institutions and the state in the process of influencing domestic values, institutions or policies. This is leading us to another question: what processes are behind the concept?

1.2 How Europeanization Works?

In order to answer how Europeanization works we have to fist answer who is the subject of Europeanization. As pointed out by Olsen (2002), there are at least five meanings of Europeanization which presuppose existence of various subjects: EU member states, candidate countries, and other subject including EU non-member states[2] or various intergovernmental or non-governmental organizations, associations, networks or individuals. It is evident that mechanism of Europeanization will work in different way for EU member state and non-EU member state. This book focus on the influence of REACH beyond EU borders, however, for analyzing such influence even some mechanisms of EU influence on EU member states might be useful.

For example Christoph Knill and Dirk Lehmkuhl (2002) distinguish three types of Europeanization which may occur in the EU member states depending on the area of Europeanization. In the areas of "positive integration" (such as environmental protection, health, consumer protection, work safety or social policy) they refer to Europeanization by institutional compliance. Simply, the EU sets standard and EU member states adopt it. This is distinctive from areas of "negative integration" related especially to common market where EU is driving its member states to abolish measures posing barriers for the free movement of goods, persons, services and capital. The key mechanism is the change in domestic opportunity structures. Finally, Knill and Lehmkuhl refers to third kind of Europeanization trough "framing domestic believes and expectations" (Knill and Lehmkuhl 2002: 259). In other words, EU promotes Europeanization in the indirect way by changing opinion of domestic actors on a cognitive basis. This mechanism is useful also for EU external dimension. While first two mechanisms are closely related to properties of the EU law (supranationality, direct effect etc.) the "Framing" perspective does not require enforceable link between EU and the state.

It is well known that there is considerable EU influence in various areas related to global commerce, notably antitrust laws, privacy regulation or food safety without EU direct interference and control. According to Anu Bradford (2012) this is caused by the ability of the EU to export its influence through institutions and standards in areas related to the EU's regulatory policies (Bradford 2012: 3).

[2] Specific sub-category among EU-non member States are except candidate countries also countries covered under EU Neighbourhood policy (see Gawrich, Melnykovska and Schweickert 2010: 1210).

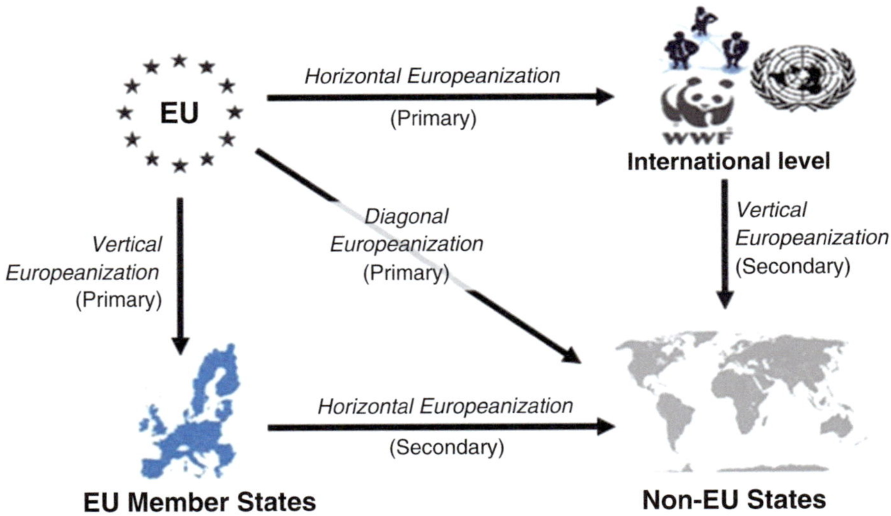

Picture 1.1 Directions of Europeanization and involved actors (Source: Author)

Europeanization may not only be a vertical process which took place only between two levels including the EU and the Member states, but as Bradford indicated, it may be more horizontal where EU influenced international actors including international organizations, institutions, networks or NGOs.

The precise distinction between vertical and horizontal influence may be not clear in the case of countries outside the EU. While between EU member state level and the EU level is the relationship predominantly vertical due to many aspects of EU membership, including acceptance of supranationality, in the case of non-EU countries the supranational link is missing. From the point of view of the non-EU state, there might be three main sources of Europeanization. First, the EU itself which is influencing the non-EU state and is as a such is the primary source of Europeanization; Second, the EU Member state which has already implemented EU law and is influencing the non-EU state. Europeanization is secondary due to existence of intermediary actor. The third source of Europeanization might be actors on the international level, such as international organizations, political networks or NGOs whose policies, norms or procedures were Europeanized and which are further influencing a non-EU state in secondary way. Picture 1.1 shows directions of Europeanization and involved actors.

It is important to note that the above picture presents certain simplification of the process. First, both terms of horizontal and vertical Europeanization are ideal types as it could be expected that there is not only single source of influence at the time. Second, in certain processes under Europeanization, the actors have broader space to pursue their own interests and defend its policies, values or procedures which are subject of Europeanization. For this reason the intensity of Europeanization vary in time as well as response to it. Third, primary and secondary Europeanization are distinguished based on the number of intermediaries providing Europeanization.

In all types of Europeanization (horizontal, vertical or diagonal) may be direct or indirect. Among EU member states direct Europeanization refers to obligation adopt EU law (norms, rules, procedures). However, Europeanization may occur also in indirect, mainly voluntary way. In relation to candidate countries direct Europeanization is connected to fulfilling EU membership requirements, however Europeanization may occur also trough indirect way by "framing" as defined by Knill and Lehmkuhl (2002: 259). In the case of non-EU countries direct Europeanization is rare as there is no enforceable obligation for those countries to follow EU norms. However, there are some exceptions such as measures taken under Common Foreign and Security Policy Mission, IPA Transition Assistance etc. EU may also directly or indirectly influence international actors including international organizations, non-governmental organizations, networks or individuals. Same is valid for secondary Europeanization which may occur in direct or indirect way, based on the relationship characteristic between transmitting and receiving actor.

The direction of Europeanization towards non-EU state is both vertical and horizontal in its nature. Verticality roots in the status of the state and the status of the EU as a *sui generis* polity. However, there is no obligation of the non-EU state to adopt EU law which is in contrast to subordination and shifting non-EU states closer to horizontal direction. This "diagonal relationship" results in diagonal Europeanization which is primary in the sense that there is missing third actor between EU and the non-member state. In reality this relationship is mainly indirect as the EU has no interest to interfere into non-EU member state domestic policies and the EU non-member state has not obligation to adopt EU rules. In other words, diagonal Europeanization is predominantly voluntary and indirect process. Further chapters will explore main attitudes to vertical and horizontal directions of Europeanization.

1.3 Vertical Dimension of Europeanization

Depending on the variables mentioned, definitions highlight very often the impact of the EU level institutions, processes or policies on a national level. This "top-down" approach has been analyzed in different theoretical perspectives focusing on *"uni-directional changes and narrow impacts"* (Radaelli 2004: 6) by conducting analysis of changes caused by EU policy implementation. A different perspective offers "bottom up" approach which starts analysis from the perspective of domestic actors. As Radaelli points out, researchers are using time and temporal causal sequences to check if, when and how EU has changed main elements in the domestic system and the interaction within (Radaelli 2004: 7). Regarding vertical direction of Europeanization, this book reflects the rather top down approach because it deals with domestic adaptation of selected non-EU countries on EU chemical policy. Claudio Radaelli (2000) suggests that the Europeanization researcher should specify what is Europeanized (domestic structures, cognitive structures, public policy…) and answer the question "how much change" has been brought by Europeanization (Radaelli 2000: 12). In line with Radaelli premises, the main focus of this book is put on the Europeanization of chemical policies and attempt to find conclusions on how much were national chemical regulations the subject of Europeanization in selected countries.

However, in general, there are according to Tanja Börzel and Thomas Risse (2000) three major dimensions where the domestic impact of Europeanization can be analyzed and traced. Those include the spheres of policy, politics and polity. While domestic policy changes reflects growing number of EU made norms the dimension of politics reflects the attempts of domestics actors to influence the policy making of EU institutions. Many works also focus on a polity dimension and the impact of Europeanization on formal and informal aspects of domestic institutions (Börzel and Thomas Risse 2000: 3–4).

For many authors Europeanization is not a one way process. For example Radaelli (2004) is referring to two terms reflecting different way of Europeanization within vertical dimension: "downloading" and "uploading" (this has nothing to do with top down and bottom up approach). Europeanization might be the consequence of "downloading" provisions of EU policy. EU institutions produces huge quantity of legal acts and some of them (directives) require transposition of EU norms into domestic law. Previously the domestic area became Europeanized when a state fulfills its obligation of transposition. It is hard to estimate, how much Europeanization occur through EU legal instruments as there are no relevant statistics and data are partial. For example, on average 84 directives are implemented in the Netherlands every year (Bovens and Yesilkagit 2010: 11) and in between 1998 and 2004 there were 750 directives implemented in Germany requiring separate implementing laws (Gaskell and Persson 2010: 38).

Transposition of EU directives might be costly if the substantial change of the system is required and new provisions are getting against existing and usual practices of the EU member states. In order to decrease costs associated with implementation of EU law states may decide to act before downloading occur and "upload" its agenda and national policy arrangements on the European level (Börzel 2002: 196). Uploading could be thus considered as preventive action before anticipated Europeanization process takes place. By uploading, member states can reduce costs because uploaded policy will better fit the current policy and as a result only little adaptation will be required. Second, uploading may reduce costs for the domestic industry which might face costs regarding adaptation to new standards. In another words, state may upload standards which are already implemented and by this step provide competitive advantage to its domestic industry. Third, uploading may help the member state to supranationalize or internationalize agenda which might be better dealt on European level (Börzel 2002: 196).

Based on its economic development and state capacities, states in response to Europeanization may adopt three different strategies. First, be more active in pushing policies at the European level to minimalise implementation costs (pace-setting);

Second, trying to block or delay policies which are costly and attempt to receive compensation (foot-dragging) or building tactical coalitions with pace-setters and foot-daggering states, which Tanja Börzel called "fence-sitting" (Börzel 2002: 197). However, central government is not the only actor who deals with Europeanization. Based on Dolowitz and Marsch (1996) Ian Bache distinguishes between voluntary and coercive Europeanization. While under voluntary Europeanization the process is welcome and supported by domestic actors, under coercive Europeanization actors oppose the changes (Bache 2003: 11). This distinction is important also in the case of REACH adaption, because such a complex and in many aspects restrictive regulation may cause concern and negative responses among subjects of regulation and thus influence not only the atmosphere or public opinion but also governments. In other words, REACH provisions could be expected in countries where domestic actors are supportive of Europeanization. This may be caused for example by the structure of chemical industry, NGOs support, the export structure and its orientation on the EU countries or the general attitude on the international level.

As already mentioned, Europeanization is often perceived as a two way interactive process between two actors (EU and the member state). However, as Burch and Gomez (2003) notes, there might be other levels with varying significance. Thus, Europeanization may occur even without the EU when for example interaction take place between EU member states in sharing best practice on some different platform (Burch and Gomez 2003: 2). They refer to "cross-loading" as a process of mutual cooperation and learning which might result in Europeanization (Hang 2011: 144). From this point of view it is important to find also other sources of international norms which may affect selected countries and for this reason we will explore international arena in the third chapter.

As already mentioned Radaelli (2000: 12) suggests that researchers should focus on the question how much change has been brought by Europeanization. This inevitably opens the question about existing space between existing legislative, values or processes and desired outcome represented by Europeanized legislative, values or processes. Tanja Börzel and Thomas Risse (2000) points out, that there must be some "misfit" between existing policies, processes or institutions and the size of misfit determines the degree of pressure for Europeanization: *"The lower the compatibility between European and domestic processes, policies and institutions, the higher the adaptation pressure"* (Börzel and Risse 2000: 5).[3] Policy misfit may occur when there are new rules established at the EU central level and those rules differ from domestic rules. This is self evident in the case of EU members, because regulations and directives are giving rights to its subjects and may be effectively enforceable by those actors, states or EU institutions. On the contrary, this is not the case of non-EU states as there is no effect of the EU law. However, in the case EU set normative example which is perceived as good practice or international standard, it could be expected that the lower compatibility with this international standard, the higher pressure for adaptation in the non-EU state. In other words the

[3] This logic is valid for the EU member states as the adaptation pressure is strengthened by institutional and legal means.

implementation of international standards matters as it is connected with the international prestige.

As Börzel and Risse (2000) points out, this misfit can be responded by three different degrees of adaptation including absorption, accommodation and transformation. Absorption requires low domestic change due to similar nature of European policies and ideas so substantial modification is not necessary. Adaptation by accommodation requires modest changes in domestic policies when existing policies are modified without changing their essential features. Transformation requires a high degree of domestic change by replacement of existing policies, processes or institutions by different ones (Börzel and Risse 2000: 5). Claudio Radaelli also adds the option of inertia where there is a complete lack of change accompanied by transposition delays or different forms of resistance towards Europeanization (Radaelli 2000: 15). The scope of Misfit can predict how the state will react on new model. However, there are some exceptions. As for example pointed out by Markus Haverland (2000) in the area of gas liberalization there was huge misfit in the Netherlands and substantial change occurred. On the other hand, in the competition policy there was minimal misfit in the Denmark and substantial change occurred as well. Similar area is hull transport in France (Haverland 2000). In order to explain this differences, some authors are going in the debate on under which conditions Europeanization occurs even further with presenting set of intervening variables.

Based on contributions by Maria Green Cowles, James Caporasso and Thomas Risse (2001) or Tanja Börzel and Thomas Risse (2003) we can add other factors such as number of institutional veto points within countries exposed to Europeanization, intermediating institutions, different power setting among actors, learning process, expectations of actors, existence of formal institutions or the nature of political culture. In other words a row of diverse aspect may influence whether REACH will be implemented or not and it might be expected that these factors will have greater influence as the link between EU and non-member states is weaker than in the case of EU member states.

Due to various and different legal cultures of selected countries and varying political systems, it could be expected that domestic response to REACH will be evolutionary, rather than revolutionary change within the system. In other words, full implementation of REACH cannot be expected due to autonomy of the political systems and limited influence of vertical direction of Europeanization. If some changes will occur, it could be expected that they will have character of partial absorption rather than accommodation.

1.4 Horizontal Dimension of Europeanization

From the geographical point of view, the majority of Europeanization studies focus on EU member states, some on the Europeanization in the candidate counties and few on the effect of Europeanization in the third countries. Frank Schimmelfennig in his review article (2012) focus on the Europeanization beyond borders of the EU. His work presents conceptual overlap of theoretical perspectives used mainly with regard to accession countries (Schimmelfenning 2012: 6). However some aspects may be well used also in application to third countries. Schimmelfenning created matrix of eight different mechanisms of the EU impact on the member states. This matrix is combining the logic of action developed in works of Diez et al. (2006) and Baurer et al. (2007) and the logic which distinguishes between direct and indirect Europeanization as presented by Schimmelfenning and Sadelmeier (2005) and Lavenex and Ucarer (2004).

Cornerstones of the matrix are distinction between "logic of consequences" and "logic of appropriateness" or between rationalism and constructivism. Under the logic of consequences actors choose the behavioural option to maximalize their utility under the circumstances. The logic of appropriateness means that actors choose the behaviour which is appropriate according to their social role and the social norms in a given situation (Schimmelfenning 2012: 7). Under the logic of consequences EU may promote Europeanization by sanctions and rewards. States are led in this case socio-economic analysis of the impacts and calculate impact of available options. Europeanization under the logic of appropriateness occurs when social learning takes place and states perceive EU rules as legitimate (Schimmelfenning 2012: 7). The second logic used in the matrix is the way of Europeanization which may occur through intergovernmental or transnational channels. In both cases Europeanization might be directly driven by the EU or in the indirect way (Schimmelfenning 2012: 8). A mix of these factors resulted in eight different modes of Europeanization presented in the Table 1.1.

Table 1.1 Mechanism of EU impact beyond the member-states

	Intergovernmental		Transnational	
	Direct	Indirect	Direct	Indirect
Logic of consequences	**(1) Conditionality**	**(2) Externalization**	**(3) Transnational incentives**	**(4) Transnational externalization**
	Intergovernmental incentives	Competition	Connective Impact	Competition
	Compulsory impact	Negative Externality		
	Compliance			
Logic of appropriateness	**(5) Socialization**	**(6) Imitation**	**(7) Transnational socialization**	**(8) Societal imitation**
	Intergovernmental social learning	Lesson-drawing	Transnational social learning	Enabling impact
	Constructive impact	Enabling impact		
	Communication	Unilateral emulation		

Source: Schimmelfenning 2012: 8

Table 1.2 Forms of primary Europeanization

		Actor's approach	
		Active/cooperative	Passive/absent
EU influence	*Direct*	Intended Europeanization	Forced Europeanization
	Indirect	Voluntary Europeanization	Latent Europeanization

Conditionality and socialization are two main modes of Europeanization. Conditionality refers to the sanctioning impact of the EU which may punish target government for hesitation to comply with the EU rules. Adoption of the EU preferred behaviour and implementation may in contrast lead to benefits such as market access, institutional ties or financial aid. Europeanization occurs directly as a consequence of the actors decision to comply with EU rules to reach incentives and avoid possible sanctions (Schimmelfenning 2012: 8).

Socialization refers to social learning when states recognize EU norms as legitimate and identify themselves with EU solutions. Other modes are considered as varieties of these two mechanisms because they may occur in more indirect or transnational way (Schimmelfenning 2012: 8). Schimmelfenning in his matrix presents several forms of Europeanization based on horizontal direction where Europeanization happens through different channels where transnational actors are involved. The above matrix is helpful in analyzing the external impact of the EU and might be applied on changes in chemical regulation caused by REACH adoption. However, despite there are eight different mechanisms of EU impact it might be impossible to identify one most suitable because logic of consequences and logic of appropriateness might run simultaneously and interact as well as direct and indirect forms of intergovernmental/transnational actions. It is thus much easier to answer the question of "what" and "to what extent" has been Europeanized than "why" and "how" it has been Europeanized. Case studies presented in this book will try to focus on these four questions. Schimmelfenning's matrix is useful for the detailed understanding of the Europeanization process, nevertheless for this book is too complex and thus does not work very well. For the purposes of the analysis conducted in this book his matrix is simplified.

Europeanization can be EU driven or domestically driven (Schimmelfenning and Sedelmeier 2005: 8) as well as direct or indirect. The way how Europeanization is driven closely reflects the attitude of a state which can be active or passive. In the case of EU non-member states (except candidate countries) we can expect that there will be no direct link between two actors as the state has no legal obligation to adopt EU rules. It can be expected that in this case Europeanization will not be EU driven. State approach and direction of Europeanization creates four different possibilities how Europeanization may take place and which are consistent with diagonal Europeanization presented above. Table 1.2 shows four forms of primary Europeanization based on EU influence and actor's approach.

When domestically driven a state plays an active role and fully cooperates with the EU to adopt EU values, rules and procedures. This may be an example of EU member states adopting EU law or of candidate countries implementing *acquis communautaire* or in the case of EU non-member states a country which is conducting domestic reform under Common Foreign and Security Policy Mission. However, state may adopt also passive role or role hostile towards EU. This may be caused for example by a political situation within the target state where eurosceptic parties lead the government or by lacking the capacities necessary for implementation. In this case Europeanization is rather EU driven and despite hesitation and the negative approach of a target state it is obliged by the law to adopt EU rules. In the case of non-compliance EU can take further legal action before European Court of Justice. This form of Europeanization could be called forced Europeanization. As in the previous example the main actors are EU member states or candidate countries during the accession process who are not fully cooperating in implementing EU law. This form of Europeanization will be rare among external non-EU actors due to state sovereignty, however it may occur in situations where EU is involved e.g. in post-conflict reconstruction.

On the other side the EU may promote Europeanization and influence on other actors in indirect ways. In this case Europeanization is not legally binding but rather a voluntary process states have undertaken for various economic, political or normative reasons. Under direct Europeanization EU plays more active role than under indirect Europeanization. The power of the EU rests in its soft-power as defined by Joseph Nye (2004). Nye claims that soft power is *"the ability to get what you want trough attraction rather than coercion or payment"* (Nye 2004: x) with special reference to culture, political ideals and policies which are seen as legitimate in the eyes of others (Nye 2004: 11). In this case the EU is actor of Europeanization just by its presence in the international relations system[4] and thus less active compared to direct forms of Europeanization.

As an actor of international relations and mixed polity *sui generis,* EU culture, values and policies represents a normative ideal which might be to various degrees reflected by other actors of international relations. States may adopt the decision to voluntary adopt these values or policies in line with rational choice theory or the constructivist approach. Reasons for "Voluntary Europeanization" might be of economic or political nature. The last form is "Latent Europeanization" which occurs when actors have a passive approach towards EU policies on the level of state but actors within the state adopt EU rules, norms and standards.

Contrary to intended or voluntary Europeanization, latent Europeanization is unintended in nature. In this case EU norms, values and procedures are not required or implemented in national law, but are followed by sub-national actors. This may be seen in the example of companies which adapted EU rules in order to have access on the EU market. These changes may cover administrative adaptation as well as changes in production processes and quality standards. Despite there being no obligation to follow EU rules in the national regulation, they voluntarily did so. The more rules implemented by sub-national actors the more latent environment for

[4] For example David Allen and Michael Smith (1990) considered Western Europe as a variable multi-dimensional presence which plays less active a role in certain areas of international interactions.

Europeanization there will be as the cost for central government to introduce EU-like regulation decreases. Until EU-like regulation is enacted, misfit remains and becomes two dimensional: intra-state misfit influencing the changes in the opportunity structure and international misfit based on the logic of framing.

This simplified matrix has several limits. For example domestic policies are not only resulting of pressure for Europeanization (see for example Goetz and Hix 2001) and other factors including national strategies, cultural and historical traditions or country specific characteristics may have effects as well. For example Markus Haverland notes that in studies there is lack of causality analysis in the interaction between European and domestic levels and those case studies are selected to deal with areas where Europeanization is clearly visible (Haverland 2005). The importance to find possible causal link and validating the impact of the EU on domestic change is stressed also by others (Ladreach 2010: 2). In case studies presented in this book it is evaluated whether changes in domestic policies are a consequence of European pressure or the causes are of different e. g. functional nature. Deficiency of sample selection is overcome in this book because Part II presents several case studies dealing with geographically and culturally different counties and as a consequence of different levels of Europeanization high variability of results is expected. Even according to Haverland comparing EU with non-EU states is a very promising strategy how to avoid the problem of bias towards EU influence (Haverland 2005: 8). Conducting several case studies then will increase relevance of the research.

Chapter 2
REACH as the Source of Europeanization

In order to respond the question "how much Europeanization?" took place in chemical legislation in selected countries it is necessary to explore how the result of Europeanization should look like or what shall the "export" from the EU look like. This chapter presents material to the core of Europeanization. Focus is given to content of Europeanization which is REACH regulation, however there follow REACH-linked regulations which are considered as integral part of EU chemical management legislation. This includes CLP regulation, Biocidal product regulation (BPR) and Prior Informed Consent regulation (PIC). Each regulation will be explored in a separate section. It is not the aim of the chapter to present the legislative in its complexity, but rather to point out the main aspects. For this reason the chapter is descriptive in its nature, first introducing the general context and then key aspects of the legislation. The four sections about EU chemical regulations (REACH, CLP, BPR and PIC) will be preceded by section introducing the situation prior REACH in order to show the contrast and revolutionary progress of existing EU legislation.

2.1 First Chemical Regulation

The pre-REACH system of chemical regulation was based on four legal acts:

- Directive 67/548/EEC – classification, packaging and labelling of dangerous substances of 27 June 1967 on the approximation of laws, regulations and administrative provisions relating to the classification, packaging and labelling of dangerous substances;
- Directive 1999/45/EC – classification, packaging and labelling of dangerous preparations of the European Parliament and of the Council of 31 May 1999 concerning the approximation of the laws, regulations and administrative provisions of the Member States relating to the classification, packaging and labelling of dangerous preparations;

O. Filipec, *REACH Beyond Borders*, DOI 10.1007/978-3-319-54154-9_2

- Council Regulation (EEC) No 793/93 of 23 March 1993 on the evaluation and control of the risks of existing substances; and
- Council Directive 76/769/EEC of 27 July on the approximation of the laws, regulations and administrative provisions of the Member States relating to restrictions on the marketing and use of certain dangerous substances and preparations.

The main problem of this regulation was that there were no provisions imposing an obligation on producers to test chemical substances. As a consequence the effects of substances on human health and environment were almost unknown due to lack of information. Before 1981 there was enlisted 100,106 chemical substances in the European Inventory of Existing Commercial Chemical Substances (EINECS) and around 70,000 in actual production, which were in majority not tested, nor risk assessment was conducted (European Commission 1998: 11; Schörlig 2004: 49, 55). There was almost no information about 30,000 of the most used chemicals and due to lack of obligation to register substances produced in quantities less than 10 tons/year the total number chemicals on the market was unknown (Heyvaert 2008: 4; Warhurst 2005: 165). The lack of information about existing chemicals was present in all categories. Some non-complex information was available to only some 3000 substances. In the case of substances produced in large quantities above 1000 tons/year there was necessary information only for 14% of substances (European Parliament 2001).

This lack of information created a very confusing situation where the health of the producing personal and even consumers was in danger. The consequences for the environment including animals, food chain or water quality was unknown. The situation was even more severe if we take into count that the effect of certain chemicals is evident only after a longer period of exposure. For example, Asbestos, DDT, CFC or polychlorinated biphenyls were in many countries allowed in use before being banned. Moreover effects may vary in time for example due to concentration used or increasing resistance of the organism. It is also important to note, that there is in reality nothing such as single exposure in a day (Santilo and Johnston 2006: 11). It is estimated that normal person is exposed to hundreds of chemicals every day (Thorton et al. 2002). For example, according to the Environmental Working Group an average US woman put on average 168 chemicals on her body in cosmetic products (ABC News 2015).

The first attempts to solve the situation within EU came by the adoption of the Council Regulation 793/93 on the evaluation and control of the risk of existing substances which placed chemicals produced in high volumes on the priority list to conduct chemical assessment. However, eight years later only 140 substances were on the list and necessary analysis has been conducted only in 11 cases (European Parliament 2001). The procedure was very slow as the placement of substance on the list lasted between 27 and 54 months (European Commission 1998: 17). This lengthy procedure was not the only problem.

The burden of proof rested on the national authority. Until the national authority received information about the harmful properties of the substance, the substance could be placed on the market without any restrictions. Also, national authorities were passive in controlling the market as there was lack of capacities and insufficient funds (European Commission 1998: 13). This system clearly put a great disadvantage on the consumer who with lack of information had to rely on passive national authorities (European Parliament 2001). Moreover, even existing norms were not fully followed by companies. For example, in the case of labelling, the European Commission conducted inspections in 100 randomly selected companies. In 25% of cases companies used incorrect classification of chemicals and in 40% of cases incorrect labelling (Commission 1998: 13).

It was evident that the existing system was ill balanced and needed substantial reform which would fill the existing information gap and strengthen the customer by shifting the burden of proof to companies. After the famous summit of Environmental Ministers in Chester, the European Commission was invited to formulate new principles of chemical regulation in its White Paper on Chemicals Policy first published in 2001. With this invitation, the battle over new regulation began in full swing.

2.2 REACH Regulation

As the abbreviation suggests REACH stands for registration, evaluation and authorisation of chemicals. However, several other aspects are not included in the abbreviation including restriction of chemicals or establishment of the European Chemicals Agency. REACH was adopted after almost a decade long period of negotiations on the 18th of December 2006 and entered into force on 1 June, 2007. The main aim of this regulation (Article 1) is to ensure protection of human health and the environment, the promotion of alternative methods for assessment of hazards of substances as well as free circulation of substances on the EU internal market while enhancing competitiveness and innovation. It is also mentioned, that the regulation is based on the principle that it is for manufacturers, importers and downstream users to ensure that they manufacture, place on the market or use such substances that do not adversely affect human health or the environment. The provisions of the regulation are underpinned by the precautionary principle (Article 1). To fulfill these ambitious aims, many implications, especially for producers and importers of chemicals exist.

2.2.1 Registration

Under registration, importers and manufacturers are obliged to register their substances, based on production volumes and hazard properties. Thus production volume determines not only the hazard properties and deadline for registration but, for example also the scope of information necessary to submit during the registration procedure. Article 23 sets the registration deadlines for so called "phase-in" substances (in general those are existing substances which under REACH enjoy extended deadlines):

- Until the 30th of November 2010 shall be registered all Carcinogenous, Mutagenic or Reprotoxic substances produced in quantities 1 ton per year or higher (CMR1 and CMR2); substances classified as very toxic to aquatic organisms which may cause long-term adverse effects in the aquatic environment (R50/53); or other phase-in substances manufactured or imported in quantities reaching 1000 tonnes or more per year per manufacturer or per importer.
- Until 31st of May 2013 shall be registered all substances produced or imported in quantities reaching 100 tonnes or more per year per manufacturer or per importer.
- Until 31st of May 2018 shall be registered all substances reaching quantities of production or import 1 tonne or more per year and per manufacturer or producer.

In order to ensure fluent adaptation on REACH, producers and importers were allowed to pre-register their substances, which give them additional time to collect necessary information (Article 28) and benefit from transition period under Article 23.

During the registration process a technical dossier (including e.g. the identity of the manufacturer and the substance, its classification and labelling, proposals for testing, study summaries etc.) and chemical safety report (including e.g. concentration limits or hazard assessment data) should be submitted (Article 10). Joint submissions by multiple registrants are also possible (Article 11). While provisions are very open to joint submission[1] and data sharing[2] (for example via Substance Information Exchange Forum), the sanction for delivering required information is absolute. Under Article 5 it is stated "no data, no market" principle. This principle made registration compulsory as substances on their own, in preparations or in articles shall not be manufactured in the Community or placed on the market unless they have been registered. Several exemptions from the general obligations to register is stated under Article 9 (the process oriented research and development falls under this provision).

A natural or legal person established outside the Community who is producer or importer of chemical substances may according to Article 8 appoint by mutual agreement the only Representative who will fulfill the duties on behalf of the importer.

---

[1] Under the new REACH-IT version (2016) joint submission are mandatory.

[2] Data Sharing provisions (especially with regard to the associated costs) have been further developed by the Commission Implementing Regulation (EU) 2016/9 of 5 January 2016 on joint submission of data and data-sharing in accordance with Regulation (EC) No 1907/2006 of the European Parliament and of the Council concerning the Registration, Evaluation, Authorisation and Restriction of Chemicals (REACH).

2.2.2 Evaluation

In general, there are three types of evaluation under REACH: examination of testing proposals submitted by registrants (Article 40), compliance check of the dossiers submitted by registrants (Article 41) and substance evaluation (Article 44). In the first two types the European Chemicals Agency evaluate information submitted by companies and in the case some information is missing or inadequate, the registrant is required to submit further information. Evaluation of chemical substances (for which stands E in REACH abbreviation) is conducted by Member States Competent Authorities who examine any information on a substance. When additional information is required, ECHA may decide to request additional information.

All substances to be evaluated are placed on a list called "Community Rolling Action Plan" which is adopted by the Member State Committee, the decision-making body within the European Chemicals Agency consisting of EU member states representatives (Article 45). The results of substance evaluation are very important as it determines further action. The result may lead to no action towards registrant where all properties of the substance proven to be safe or sufficiently under control with the measures already taken. However in opposite cases the action may result in the EU proposal to widen risk management including restrictions, identification of substance as a substance of very high concern, harmonized classification, imposition of voluntary measures etc.

2.2.3 Authorisation

The Authorisation process should ensure good functioning of the EU internal market and assure that risk from substances of very high concern are properly controlled and that these substances are progressively replaced by alternative substances where safer alternative is available and these are economically and technically viable (Article 55). Substances that have been not authorised or are not subject of an exception (such as research or development purposes) shall not be placed on the market (Article 56). Article 57 states the scope of the authorisation, including carcinogenic substances, mutagenic substances, substances toxic for reproduction, substances which are persistent, bio accumulative and toxic, very persistent and very bio accumulative or under certain conditions also endocrine disruptors.

The authorization process consists of two phases. In the first step substances which may cause serious effects on human health or the environment are identified (Article 59). European Chemical Agency or any EU member state may propose to the European Commission that the substance should be identified as a Substance of Very High Concern and thus placed among substances eligible for authorisation (Annex XIV REACH). In the second phase such identified substance is placed on a public Candidate list so the industry and stakeholders may deliver information relevant, for conducting socio-economic analysis which, together with other information about the substance, will serve as a basis for decision whether include identified substance on Authorisation list or not.

However, if the producer, importer or downstream user will prove, that risks related to the use of such substance are "adequately controlled", it may apply for authorisation. Even when this is not proven, authorisation may be granted if socio-economic benefit outweighs the risks and that there is no suitable alternative. If applicant is not satisfied with the agency decision, a review proposal may be submitted within 18 months decision taken place. In the case of new information about risks, available alternatives or socio-economic impact review process may be whenever open (Article 61).

An important part of the REACH system is the restriction procedure despite it not being included in the REACH abbreviation. The restriction procedure is an important tool to protect human health and the environment. When there is a suspicion that certain substance posse an unacceptable risk, the EU member state, European Chemicals Agency or European Commission may introduce the restriction procedure. As well as in the case of authorisation, the proposal for restriction is being made public for relevant comments prior decision take place (Article 68). Such proposal contains information about identified risks, availability of alternatives or the costs and benefits associated with the restriction. The final decision is made by the European Commission who is by the opinion informed by European Chemicals Agency. It is important to note, that restriction of hazardous chemicals was one of the consensual aspects of chemical regulation within the EU which builds on previous existing legislation which has also a normative impact outside the EU. The Council Directive (EC) 2002/95 of 27 January 2003 on the restriction of the use of certain hazardous substances in electrical and electronic equipment soon became the global standard in the restriction of six hazardous chemicals included within the electronic device.

2.2.4 Establishment of the European Chemicals Agency

Under Title X of REACH the European Chemicals Agency was established to carry out the technical, scientific and administrative aspects of REACH. The Helsinki based Agency is the driving force among regulatory authorities in implementing the EU chemical legislation which recently has grown. Today ECHA has responsibilities under:

- REACH regulation (2007)[3]
- CLP regulation (2009)[4]

[3] Regulation (EC) No 1907/2006 of the European Parliament and of the Council of 18 December 2006 concerning the Registration, Evaluation, Authorisation and Restriction of Chemicals (REACH) establishing a European Chemicals Agency, amending Directive 1999/45/EC and repealing Council Regulation (EEC) No 793/93 and Commission Regulation (EC) No 1488/94 as well as Council Directive 76/769/EEC and Commission Directives 91/155/EEC, 93/67/EEC, 93/105/EC and 2000/21/EC.

[4] Regulation (EC) No 1272/2008 of the European Parliament and of the Council of 16 December 2008 on classification, labelling and packaging of substances and mixtures, amending and

- Biocidal Products regulation (2013)[5]
- Prior Informed Consent regulation (2014)[6]

The growing tasks and responsibilities of the agency require sufficient institutional structure. Article 76 REACH mentions Institutions of the agency which includes Management Board, an Executive Director, the Secretariat, Board of Appeal, and Forum for Exchange of Information on Enforcement and three Committees: a Committee for Risk Assessment, a Committee for Socio-economic Analysis and a Member State Committee. These Committees and Forum are entitled to establish relevant working groups and seek scientific or ethical information from appropriate sources. While Committee for Risk Assessment and the Committee for Socio-economic Analysis are responsible for preparing scientific opinions in the various phases of evaluation, authorisation or restriction, the Member State Committee is resolving potential divergences of opinions proposed by the Agency regarding substances of very high concern or other issues. Similarly to other international organizations the Secretariat works under the leadership of the Executive Director and provides technical, administrative or scientific support to other bodies, including administration of pre-registration, registration, evaluation, database maintenance or information provisions. Board of Appeal decides appeals against decision taken by the Agency.

The structure of the agency and its bodies are much more similar to international organization rather than national agency or ministry. Its position within the regulatory system is unique, because agency servers as central hub for 28 national authorities and compared to other EU agencies has high level of discretion (see Enhert 2008). The success of ECHA leads to increasing tasks of the agency in other areas of chemical regulation.

2.3 CLP Regulation

The CLP regulation entered into force in January 2009 with the aim to ensure that hazards presented by chemicals will be well understood by workers and downstream users. In some aspects it is a complementary regulation to REACH which

repealing Directives 67/548/EEC and 1999/45/EC, and amending Regulation (EC) No 1907/2006 (abbreviated as CLP).

[5] Regulation (EU) No 528/2012 of the European Parliament and of the Council of 22 May 2012 concerning the making available on the market and use of biocidal products (abbreviated as BPR).

[6] Regulation (EU) No 649/2012 of the European Parliament and of the Council of 4 July 2012 concerning the export and import of hazardous chemicals.

replaced Dangerous Substances Directive[7] and Dangerous Preparations Directive[8] and aligns EU legislation with Globally Harmonised System of Classification and Labelling of Chemicals (GHS). The new complex regulation updated common rules for classification and labelling of substances so the hazard can be recognized by the workers and users before the chemical will be handled. Duties under CLP are divided according to supply chain position including the manufacturer and importer of substances or mixtures, producer of specific articles, downstream users or distributors. If a substance is placed on the EU market, according to Article 40 CLP the agency should be notified within one month about classification and used labelling.

Classification is made in four steps defined in Chapter 1 of CLP. First, available data about the substance or mixture are collected (Article 5), then all information is evaluated for adequacy and reliability (Article 7, 8), third, review of the information against the classification criteria follows (Articles 9 to 11) and finally the decision on classification is taken. According to Article 13 CLP when a substance is classified as hazardous or a mixture contains one or more hazardous substances above a certain threshold, the substance must be properly labelled. Such labelling includes supplier identification details, the nominal quantity of a substance or mixture in the packages made available to the public, product identifies and hazard pictograms, signal words, hazard statements and other information mentioned in the Article 17.

The only exception from these rules is small packages (less than 125 ml) which are difficult to label. Hazardous substances or mixtures supplied to public must have child-resistant fastenings and tactile warnings (Article 29). CLP covers also the labelling during transportation. If the labelling or marking is in accordance with transport legislation, it is sufficient when outer packaging of a hazardous substance is subject to both the Transport and the CLP rules. In this situation the CLP labelling does not need to appear.

If substance or mixture is classified as hazardous according to CLP, or belongs to BPT or vPvB substances, or is identified as a substance of very high concern, then the supplier shall provide Safety Data Sheet[9] with accordance with Annex II of REACH.

In order to protect the information about full composition of a mixture on the label or in the safety data sheet, a supplier which is concerned about possible business information disclosure may request in line with Article 24 the use of alternative chemical name of the mixture (Article 42). In order to make classification and labelling easier ECHA maintains CLP inventory where all notified and registered

[7] Council Directive 67/548/EEC of 27 June 1967 on the approximation of laws, regulations and administrative provisions relating to the classification, packaging and labelling of dangerous substances.

[8] Directive 1999/45/EC of the European Parliament and of the Council of 31 May 1999 concerning the approximation of the laws, regulations and administrative provisions of the Member States relating to the classification, packaging and labelling of dangerous preparations.

[9] Safety data sheets (SDS) are tool for ensuring that suppliers provide information sufficient for safe use of their substances and mixtures within the supply chain.

Picture 2.1 Hazard pictograms under CLP/GHS (Source: Author, based on CLP)

substances are listed. User may find there the name in the IUPAC Nomenclature, the name of the substance given in EINECS and the classification and the labelling of the substance (Article 18). Picture 2.1 shows pictograms used in hazard labelling.

2.4 Biocidal Product Regulation

The Biocidal product regulation (BPR) entered into the force on 1 September 2013 and replaced existing Biocides Directive,[10] which for the first time defined biocidal products as: *"Active substances and preparations containing one or more active substances, put up in the form in which they are supplied to the user, intended to destroy, deter, render harmless, prevent the action of, or otherwise exert a controlling effect on any harmful organism by chemical or biological means"* (Article 2). Biocidal products are used to protect humans, animals, plants or materials against

[10] Directive 98/8/EC of the European Parliament and of the Council of 16 February 1998 concerning the placing of biocidal products on the market.

harmful organisms including insect or bacteria by the chemical or biological action of the substance contained in biocidal product. In particular, a biocide could be for example a disinfectant, rats poison, repellents against insect, water purification tablets etc. A product with similar effects might be dangerous to health of environment and thus EU with the BPR aims to ensure high level protection. Therefore, placement of biocidal products on the market is a subject of authorisation procedure (Article 17) which is central part of the regulation. However, the regulation is much more complex as it distinguishes between approval of active substances and authorisation of products. While active substance approval is made on the EU level, the approval of product placement is made on the Member state level.

2.4.1 *Existing and New Substances*

Regulation distinguishes between new active substances and existing active substances. Existing chemical substances which were placed on the market before 14 May 2000 may be still be placed on the market, however are subject of the Review Programme[11] which is foreseen to be completed in 2024. The Review Programme sets out rules for systematic examination of all existing active substances.

The European Commission keeps the database of approved active substances called the Union list of approved active substances (Article 9). Next to this database, Commission decisions on approved and non-approved substances are published in the Official Journal of the European Union. All active substances used in this list are considered safe and might be used. So before applicant requests authorisation to place a product on the EU market, it is necessary that this product contains active substance listed in the Union list of approved active substances. The aim of active substances approval is that substances with the worst hazard properties should be excluded from the market except situation where non-approval will lead to disproportionate negative impact for the society.

In the case the substance is not listed, the applicant may apply to be included. Applicant who wants the new active substance to be placed on the list sends an application to the agency, informing it of the name of the competent authority of the Member state it proposes to evaluate the application. Application is evaluated according to article 8 by the competent authority within one year period. The member state competent authority forwards the results of the evaluation to the ECHA Biocidal Products Committee which creates an opinion within 270 days. This opinion later serves as the basis for the decision of the European Commission about approval of the new active substance. Regulation sets out also the rules for substitu-

[11] Commission Delegated Regulation (EU) No 1062/2014 of 4 August 2014 on the work programme for the systematic examination of all existing active substances contained in biocidal products referred to in Regulation (EU) No 528/2012 of the European Parliament and of the Council.

tion (Article 10). As approval is granted for limited time period of 10 years maximum, regulation sets conditions for renewal or possible review of an active substance. In both cases application is evaluated by competent authority (Article 14) and the decision is subject of review under Article 15. In general, only safe active substances meeting conditions for approval (Article 4) and not falling under exclusion criteria under Article 5 are to be approved (such as CMR, PBT and vPvB substances, endocrine disruptors and similar).

2.4.2 Authorisation of Products

For producers or importers who wants to place their products on the market there are three ways how to get authorisation depending on the product and number of countries they want place the product. First possibility is to use national authorisation (Article 29) and mutual recognition of authorisation (Article 32). Second possibility is to apply for Union authorisation (Article 41) and third, for the substances which are apparently safe and fulfills five cumulative criteria including that it does not contain substances of concern, nanomaterials or do not require personal protective equipment, a simplified procedure is available. This procedure allows placing product within all member states (Article 25).

2.5 Prior Informed Consent Legislation

Prior Informed Consent (PIC) legislation has been approved to regulate export and import of hazardous chemicals, including industrial chemicals, pesticides and biocides that are listed in Annex 1. The aim of this regulation is to promote shared responsibility and cooperation in the international movement of hazardous chemicals in order to protect human health and the environment from the potential harm and contribute to the environmentally sound use of hazardous chemicals (Article 1).

This regulation deals with chemicals banned or restricted in their use for health or environmental reasons. Such chemicals include for example atrazine, mercury, ethylene oxide, benzene or chloroform. If a company wishes to export such chemicals outside the EU, the export is subject of two requirements: export notification and explicit consent. In some aspects PIC covers also import of chemicals, however majority of import is covered under REACH and Biocidal Products Regulation (ECHA 2015: 6). PIC implements in the EU the Rotterdam Convention on the Prior Informed Consent Procedure for Certain Hazardous Chemicals and Pesticides in International Trade, which is introduced in the following chapter among other international instruments. However, it is important to note that EU implemented additional provisions in the PIC regulation going beyond requirements of Rotterdam convention (Article 5), such as the requirements for export notification and explicit consent to all countries, not only countries covered under the convention. Moreover, the intended use of the chemical in the target country is not considered (ECHA 2015: 7).

PIC does not apply to chemicals which are imported or export for the purpose of research or analysis in the quantities unlikely to cause serious health or environmental damage. For the purposes of PIC, the safe quantity is considered to be less than 10 kg of chemical per producers and calendar year (Article 2). The obligations for producer vary according to groups under Annex 1. For example part 1 of Annex 1 covers industrial chemicals for professional or consumer use, pesticides used as plant protection products or other pesticides and biocides that are banned or severely restricted in the EU. All such chemicals require export notification procedure. Chemicals under part 2 require an additional statement, which contains the declaration of the importing country, and that the authorities agree that such chemicals will be imported. Under part 3 all entries requires export notification and explicit consent, which is published in the PIC circular under the Rotterdam Convention (updated twice a year) with the condition that other criteria are met.

2.5.1 Export Notification Procedure

In the EU, the exporter shall notify its intention to export certain chemical under Annex I, part 1, to the designated national authority. Each notification will receive a unique reference identification number which will be used during whole export process including customs control. Export notification requires many data including identity of the substance, information on the exporter and export, information on precautions, summary of physic-chemical and toxicological properties, uses of chemicals etc. There are three important deadlines within the export procedure, involving the exporter, the European Chemicals Agency and the designated national authority in the non-EU state. The exporter should at least 35 days before the export is due to take place notify the designated national authority; 25 days before the export is due to take place the national authority validates the notification and forwards it to ECHA; and 15 days before the export is due to take place the ECHA sends the notification to the non-EU designated national authority in the importing country (Article 8).

2.5.2 Explicit Consent

For the chemicals listed in Annex 1, part 2 and 3 additional procedure including explicit consent is required (Article 14). In two particular situations a waiver may be granted. First, when all requirements has been met and the designated authority in the non-EU state has not responded within 60 days and second, when notified chemical is listed in part 2 of Annex I and is exported to the OECD country, where the use of such chemical is allowed and the chemical is licensed (Article 14). An explicit consent is valid for three years (if not stated different expiration date in the consent). Any other company may export same chemical within this period to the country giving consent (Article 14). For this purpose ECHA keeps a dataset of valid and expired consents (Article 6).

Chapter 3
Global Sources of Influence

European Union law is not the only source of influence and inspiration in the international arena regarding chemical management. There are several important tools and initiatives regarding chemical management on the global level. The main aim of this chapter is not to comprehensively explore all conventions, treaties and initiatives dealing with chemicals. This is not possible due to large area covered. This chapter rather introduces the main aspects of the most important tools including the Basel Convention, Rotterdam Convention and Stockholm Convention which are introduced in separate chapters. Other important instruments and initiatives, such as the Aarhus Protocol, Responsible Care or Safe Planet are introduced as well. This chapter is descriptive in its nature, exploring the material nature of chemical regulation at the international level which might be, next to the EU, other source of influence on the countries outside the EU. It helps to distinguish between Europeanization and globalization as it presents instruments on the global level.

It is important to note that presented tools and initiatives have been selected according to their importance and this process was partly arbitrary. First, the nature and content of conventions vary. While some are more local or declaratory, others might be global and more binding. Second, the area of chemical management is not clearly marked. Does it cover also waste management and air or water pollution? For this reason only the most important conventions and initiatives are explored, however other might be mentioned without specific focus.

3.1 Basel Convention (1992)

In the 1970s and 1980s the Western world experienced increased environmental awareness among its citizens, including public resistance to the disposal of hazardous wastes. For this reason companies were looking for cheap waste disposal in the countries where public awareness was weak and regulations non-existent or minimal (mainly affecting Africa and other developing countries). There are several

O. Filipec, *REACH Beyond Borders*, DOI 10.1007/978-3-319-54154-9_3

cases which were discovered and fall under the practices described as "toxic colonialism". For example Zada Lipman described the Koko Incident of 1987 where several thousand tons of toxic and radioactive waste was exported from Italy to the city of Koko in Nigeria. In another case Guinea-Bissau was offered 600 USD to dispose 15 million tons of toxic waste. As pointed out by Zada Lipman, similar contracts were reported in countries such as Namibia, Guinea, Sierra Leone or Haiti (Lipman 2002: 67). Very famous is also the story of the Khian Sea ship, which in 1986 was loaded with 14,000 tons of incinerator ash from Philadelphia. The ship was supposed to transport the toxic waste to New Jersey, however New Jersey banned the import in 1984 due to high concentrations of arsenic, cadmium, mercury, dioxins and other toxic substances (Benton-Short and Benton 2008: 195).

Over the next 14 months the ship sailed around the world trying to unload the cargo in other countries, including the Dominican Republic, Honduras, Panama, Guinea Bissau or Antilles. In 1987 the Haitian government issued a permit to import fertilizers which were used by the crew to dump 4000 tons of toxic waste on the beach near the city of Gonaives before being interrupted by Greenpeace following the cancellation of permission. The ship continued its journey to other countries, changing names and registration. *"Two years, three names, four continents, and 11 countries later, the troublesome cargo was still on the board. Then, somewhere in the Indian Ocean between Singapore and Sri Lanka all the ash disappeared"* (Cunningham and Cunningham 2015). This and many other practices of toxic colonialism contributed to the creation of the Basel Convention.

The Basel Convention (in full The Basel Convention on the Control of Transboundray Movements of Hazardous Wastes and their Disposal) was negotiated and later adopted on 22 March 1989. Basel Convention entered into force on 5 May 1992. From 53 signatories the number of parties increased to 183 (as of September 2015), including the European Union. Haiti and USA have signed the convention but not yet ratified (Basel Convention 2015b).

In Article 4 of the convention Parties expressed the obligation to prohibit the import and export of hazardous waste to other parties of the convention in the cases that the importing country is not giving written consent for the specific import. Moreover, states shall adopt necessary measures to prevent persons within their jurisdiction to transporting or disposing hazardous wastes beyond borders. Thus, transboundary movement of hazardous wastes should be reduced to minimum. The Convention sets out the rules for transboundary movement between parties (Article 6), including notification procedure and written consent. For the proper functioning of the convention a Conference of state parties has been established (Article 15) with Technical Secretariat (Article 16). Possible disputes between parties are submitted to the International Court of Justice (Article 20). The Convention has several annexes including Annex I, which presents types of wastes to be controlled; Annex II presents categories of wastes requiring special consideration, such as wastes collected from households or residues arising from incineration of such wastes; Annex III introduces list of hazardous characteristics; Annex IV presents disposal operations; Annex V sets out information to be provided on notification and movement documents; or Annex VI deals with the rules for arbitration.

The adoption of the Convention was for many countries disappointing as the transport of hazardous wastes was not banned completely. The only result was the establishment of export notification and written consent which was later known as "Prior Informed Consent". This disappointment from the Basel Convention led in 1995 to the Basel Ban Amendment, which prohibits shipment of hazardous waste from developed countries to developing countries. Despite the amendment being accepted by 79 states it has not entered into the force because it requires the ratification of ¾ countries to the Convention. However, some states object that not the ratification of ¾ of Convention signatories is required, but ¾ of the countries signed under the protocol (Basel Convention 2015b). Some states were not satisfied with the current status and were still concerned about the possibility of shipping hazardous wastes to the countries unable to handle it. The EU banned waste shipment in its European Waste Shipment Regulation[1] (2006) and Switzerland launched with Indonesia a so called "Country-led Initiative" (CLI) to improve the effectiveness of the Basel Convention. The consultations were successful, resulting in the decision adopted at the Conference of the State Parties (2011) to take actions in improving the effectiveness of the Basel Convention, including facilitation of entry into force the Ban Amendment, developing guidelines for environmentally sound management (ESM) and improving legal clarity of the Convention. The Conference of the Parties in 2013 appreciated the work of the CLI (Decision BC-11/1). The Basel Convention remains an instrument which needs to be strengthened. Its weakness led to diverse reactions. While some actors adopted more extensive regulation (EU) or focused more on an international approach (Switzerland and Indonesia), especially African states moved to closer regional cooperation in the early 1990s.

In January 1991 twelve states of the Organization of African Unity signed in Mali the Bamako Convention (in full: Bamako Convention on the ban on the Import into Africa and the Control of Transboundary Movement and Management of Hazardous Wastes within Africa). The Bamako Convention has similar language as the Basel Convention; however the provisions are much stronger, imposing a total ban on the import of hazardous wastes. In comparison to the Basel Convention, Bamako Convention in its Article 2 extended the scope of the convention to radioactive wastes. As Bamako Convention entered into the force on 22 April 1998 and as of 2013 out of 54 African countries there are 35 signatories and 25 countries already ratified the convention (AU 2013).

Very similar to the Bamako Convention is its Pacific "sister" called the Waigani Convention. The Waigani Convention was signed in 1995 in Papua New Guinea and entered into the force in 2001. In comparison with Basel, its effect is limited to the Pacific Forum region and thus less states decides about its development. While the scope of the Basel Convention covers only 12 nautical miles of the territorial sea, the Waigani Convention covers also the Economic Exclusion Zone of 200 nautical miles (Article 1). Similarly, to the Bamako Convention, it includes radioactive wastes (Article 4).

---

[1] Regulation (EC) No 1013/2006 of the European Parliament and of the Council of 14 June 2006 on shipments of waste.

Both Bamako and Waigani conventions has proven that regional based cooperation can strengthen some provisions which are problematic under conventions with many state parties. These local conventions seem to be more flexible as the number of parties is reduced.

3.2 Rotterdam Convention (2004)

The Rotterdam Convention on the Prior Informed Consent Procedure for Certain Hazardous Chemicals and Pesticides in International Trade was signed on 10 September 1998 and entered into force on 24 February 2004, however the idea has been developed since the mid 1980s under voluntary information exchange programmes administered by UNEP and FAO. The FAO introduced the International Code of Conduct on the Distribution and Use of Pesticides (1985), which presented voluntary standards of conduct for all public and private entities in the distribution and the use of pesticides, especially in the countries where the regulation was inadequate (FAO 2015). Two years later the Governing Council of UNEP adopted the London Guidelines for the Exchange of Information on Chemicals in International Trade (UNEP 1989). In 1989 the two organizations introduced the Prior Informed Consent procedure on a voluntary basis.

The important impulse came from the Rio Earth Summit which took place in 1992 in Brazil. In its conclusions it called for the legally binding instrument of PIC procedure by the year 2000. Following this decision the FAO Council and UNEP mandated their Executive Heads to open negotiations which started in 1996 and were concluded in March 1998 where the final text was for the first time presented (Rotterdam Convention 2015a). Within a decade, voluntary principles found their place in a legally binding international agreement.

Similarly to 1980s initiatives, the Rotterdam Convention aims to promote shared responsibility and cooperation among its Parties in the field of international trade of certain hazardous chemicals and to contribute to environmentally sound use of such chemicals, including information exchange about their characteristics (Article 1). The Convention applies to banned or severely restricted chemicals and severely hazardous pesticide formulations. It explicitly stated that it does not apply to wastes (Article 3). Chemicals which are subject of PIC are listed in Annex III. There are two mechanisms under the Rotterdam Convention. The PIC Procedure is a mechanism for formally obtaining and dissemination the decisions of importing Parties as to whether they wish to receive the shipment of chemicals listed in Annex III. In this case a Decision Guidance Document is prepared for a state party to consider the risk connected with the handling and use of chemicals. Each state party is required to take the decision to whether or not it wishes to import chemicals listed in Annex III. This decision is sent to the Secretariat which forwards the information to other parties via the PIC Circular where all responses are available (Article 6). In other words all state parties know what chemicals are accepted by other parties and what are not allowed to be imported. Under the information exchange procedure all parties to the convention are required to notify other parties directly or via the Secretariat about regulatory measures taken to ban or restrict hazardous chemicals. However, information exchange covers also possible information about problems caused by hazardous chemicals, future plans to export banned chemicals within its territory, information about safety data sheets and its distribution or labelling requirements for exports under PIC procedure (Article 14).

Theoretically, the Convention increases the information advantage among its parties. However, the necessity to act is not on the side of each member state. In the case of non-compliance there are no sanctions for the party of the convention and despite this mechanism is still under consideration by the Conference of the Parties (Rotterdam Convention 2015b), the absence of such mechanism makes the Convention something merely than a voluntary obligation.

3.3 Stockholm Convention (2004)

In the 1990s the international community had to deal with another environmental challenge caused by Persistent Organic Pollutants (POPs).[2] Before the discovery of the negative effects of Persistent Organic Pollutants were widely used as pesticides, insecticides, solvents or pharmaceuticals. The most infamous of such chemicals is DDT which has been used as an insecticide to protect against malaria and typhus. After Second World War DDT became the predominant method of insect control. It was inexpensive, easy to produce, very effective and at first sight harmless to people. Between 1955 and 1969 million gallons of DDT were used to annihilate the mosquito population resulting in a considerable decrease in malaria (Kinkela 2011: 1). As the use of DDT increased the first reports about possible negative effects appeared (Gunter et al. 1998: 198). The publication of the book Silent Spring by Rachel Carson increased awareness about DDT impacts, resulting in public outcry that was reflected in the 1972 DDT ban in the US. In the 1970s and 1980s the ban of DDT spread to other countries. The ban of DDT was controversial: while analysis proved that DDT is very persistent and carcinogenous, critics of the ban points out that the increase of malaria and other insect related diseases killed more people. For example, Robert Gwadz from the US National Institute of Health says that the DDT ban may have killed 20 million children (Gwadz, cited in Finkel 2007).

Despite the controversy, DDT and other 11 chemicals known collectively as the "Dirty Dozen Chemicals" became subject of the Stockholm Convention, which was negotiated between June 1998 and December 2000. The Stockholm Convention was signed on 22 May 2001 in Sweden and entered into the force on 17 May 2004, having 179 Parties (Stockholm Convention 2015a) which are dedicated to protection of human health and the environment from POPs (Article 1). The obligations related

---

[2] POPs are substances which are resistant to environmental degradation. Because of this resistance they bio accumulate and thus may have significant impact on human health and environment. They may cause cancer, hormonal dysfunctions or be toxic to reproductive health.

to chemicals covered by the convention vary according to annex. For example Annex A (elimination) lists chemicals which should be prohibited and/or eliminated in their production and use, as well as the import and export (Article 3). Chemicals listed under Annex B are restricted in the production and use, as well as the import and export (Article 3). Annex C enlisted chemicals where unintentional release is subject of reduction or elimination (Article 3). States have also the obligation to ensure that stockpiles and wastes related to POPS are safely managed (Article 6).

It is important to note that Annexes have several times been amended and new chemicals were included. Each state party may submit proposals for amendments to annexes to the Secretariat (Article 8) which is considered by the POPs Review Committee and in the case of positive evaluation submitted to the Conference of the Parties for decision. This has been done several times since the convention entered into the force. For example, in May 2009 the Conference of the Parties extended Annexes to include nine new POPs. In 2011 Endosulfan was added to Annex A and in 2013 the same annex was extended to include hexabromocyclododecane (Stockholm Convention 2015b). Parties of the convention are also encouraged (Article 9) to information exchange, including information on the reduction or elimination of the production, use and release of POPs or information on alternatives to POPs.

The Stockholm Convention presents an important step in the global fight against hazardous chemicals. However, the issue of POPs is much more complex due to persistency. POPs may travel between continents when they evaporate from water or land surface into the air, later are moved by winds and finally can return to Earth on particles in snow or rain. As the US Environmental Protection Agency points out, POPs are transboundary travelers which have been found thousand miles away from their source. They travel in oceans, rivers, lakes or with the help of migratory species (EPA 2015a). For this reason it is necessary to eliminate all sources of POPs and make the Stockholm Convention universal.

The issue of air pollution has been addressed by the 1979 Geneva Convention on Long-range Transboundary Air Pollution which entered into force in 1983 and since then has been extended by eight protocols. According to Article 2 states shall, as far as possible, gradually reduce and prevent air pollution including long-range transboundary air pollution and shall exchange relevant information (Article 3) and review their policies (Article 4) in the light of purpose of the convention. POPs are addressed in the 1998 Aarhus Protocol, which entered into force in 2003 and was amended in 2009. The original protocol focuses on a list of 16 substances with the objective to eliminate any discharges, emissions and losses of POPs. While some are subject of the production ban, others were scheduled for elimination (UNECE 2015). The amendment of the protocol from 2009 added another seven substances and revised some obligation on certain substances, including DDT.

The 1998 Aarhus Protocol finished at the time when the Stockholm Convention negotiations were in the beginning and entered into the force sooner. In some aspects regarding the ban and restriction of certain POPs both documents are the same. However, while the Stockholm Convention is almost universal with its coverage, the Geneva Convention on Long-range Transboundary Air Pollution covers only 51 parties, including the USA, Canada, European countries, Russia and some post-Soviet counties. Latin America, Africa, Australia, Pacific Region and most of the Asia are not covered (UNECE 2015). The Stockholm Convention raised the issue of POPs on a global level.

Today, the three conventions (Basel, Rotterdam and Stockholm) build the cornerstone of hazardous chemicals management on a global level. Parties to the conventions were aware of additional value which might bring synergy of three conventions and have taken series of decisions to create synergies process that aims to strengthen implementation of all three conventions including the reform of secretariats on administrative and operational level. In 2010 the first simultaneous extraordinary meeting of the Conference of the Parties took place and adopted decisions regarding joint activities, joint managerial function, joint services, the synchronization of budget cycles, joint audits and review arrangements. Later on, the conferences adopted substantively identical decisions to further cooperation and the Executive Secretary of the Basel, Rotterdam and Stockholm conventions has been requested to prepare proposal for the organization of the secretariats (Synergies 2015). In 2013 Conferences of the Parties adopted an "omnibus decision" on enhancing cooperation and coordination among the three conventions and state parties endorsed the matrix-based management approach with the organization of the Secretariat. Since 2015 there are harmonized decisions in a number of joint activities, including technical assistance or regional centres (Synergies 2015). It is important to note that the synergy has its economic and political implications such as more cost-effective implementation, integrated approach towards chemical and wastes management, better coordination in technical assistance, improved management of available resources and raised profile of the issue (Synergies 2015).

In the connection with synergy it is important to mention the Safe Planet Campaign. This campaign was launched during first simultaneous extraordinary meeting in February 2010, is run by UNEP and FAO in order to increase public awareness about hazardous chemicals and wastes. Since then Safe Planet has attracted many partners around the world and spread the activities and collaboration in areas of arts, culture, sports, science, education and other fields (UNEP 2010). All the activities of Safe Planet help to increase awareness about hazardous waste and chemicals and represent important dimension of the synergy.

The synergy of three systems seems to be another step forward in improving global management of hazardous chemicals and as such is an important source of norms and values influencing national law. Conventions created unique international systems which is accepted by the majority of states and which presents global standard in hazardous chemicals management.

Table 3.1 shows participation of selected countries (entry into force) in above three mentioned conventions and GHS, which is further explored in the next chapter. The USA and Malaysia seems to be the most reluctant states regarding implementation of the Conventions while Russia and India in the implementation of GHS. As noted earlier, all Conventions and the GHS have been implemented in all 28 EU member states.

Table 3.1 Participation of selected states in the global instruments

State	Basel convention	Rotterdam convention	Stockholm convention	GHS
USA	(1996)	x	x	I
Brazil	1992	2004	2004	I
Russia	1995	2011 (a)	2011	V
India	1992	2005 (a)	2006	Na
China	1992	2005	2004	I
Japan	1993	2004	2004	I
South Korea	1994	2003	2007	I
Malaysia	1994	2002 (a)	x	I
Indonesia	1993	2013	2009	I
Singapore	1996	2005 (a)	2005	I
Thailand	1998	2002 (a)	2005	I
Philippines	1994	2006	2004	I-
Canada	1992	2002 (a)	2004	I
Australia	1992	2004	2004	I
New Zealand	1995	2003	2004	I
Switzerland	1992	2002	2004	I

I implemented, *I-* not fullfy implemented, *V* in the case of Russia means "Voluntary"
Normal text means ratification, (a) means accession
Source: Author, based on Basel, Stockholm and Rotterdam Conventions (2015)

3.4 Other Important Tools and Initiatives

It is important to know that there are other global, or at least international initiatives which deals with chemical management. While some were developed by international organizations other comes from public sector.

Under the United Nations, the Globally Harmonized System of Classification and Labelling of Chemicals (GHS) were developed. The roots of this system may be tracked back to the 1992 United Nations Conference on Environment and Development (UNCED) which provided international mandate. In the following years the work has been done with involvement of many international organisations, including the International Labour Organization, the Organization for Economic Cooperation and Development (OECD) and the UN Economic and Social Council. The system was completed in 2001 and the United Nations Economic and Social Council's Sub-Committee of Experts on the Globally Harmonized System of Classification and Labelling of Chemicals (UNSCEGHS) is responsible for maintaining the GHS (UNECE 2011: iii). Since then the GHS system is biannually updated.

In general, GHS is a system which allows to communicate some data about properties of chemicals. Before GHS, there were national systems for labelling and information sharing about chemicals, resulting in many different standards which varied significantly in the scope of information communicated to downstream users. GHS is establishing a system for hazard classification (including physical, health and environmental hazards) and the hazard communication in the form of labels with united visual identification. It also sets requirements for the safety data sheets.

As with other tools GHS also requires implementation. The deadlines vary according to the state and the provisions for substances and mixtures. While some countries have already implemented GHS in its national law (see provisos chapter about CLP about status in the European Union), in some countries implementation is expected. According to the server ChemSafetyPro, full implementation (as of 2015) has been done in the EU, USA, Switzerland, China, Brazil, South Korea, Japan, Malaysia, Philippines, New Zeeland. Other countries have set their national deadlines: Turkey (2016) , Taiwan (2016), Singapore (2016), Indonesia (2016), Vietnam (2016), Australia (2017), Thailand (2017) and some other countries consider implementation, notably Hong Kong, Israel, Russia, South Africa, Argentina, Mexico, India and Pakistan (ChemSafetyPro 2015a, b).

GHS has an aspiration to become the global standard. However, while USA, EU, Australia and Asia seems to be already involved in this system, Africa and some other low-income states are missing. In the following chapter we will look at the reasons why some countries are interested in GHS and harmonise standards while some not. The same can be said about the implementation of International Labour Organisation Convention No. 170 of 1990.

The convention concerning Safety in the use of Chemicals at Work was signed in 1990 and entered into force in November 1993. The main subject of this Convention is to improve occupational safety and health of workers working with chemicals. The scope of the Convention (Article 1) applies to all branches of economic activity in which chemicals are used. Despite extensive scope, there are many exceptions when a Member of the Convention may exclude particular branches out of scope including "special problems of substantial nature" (Article 1) which might in reality cover many possible problems. The convention sets minimal requirements for classification systems (Article 6), labelling and marking (Article 7), chemical safety data sheets (Article 8), responsibilities of suppliers (Article 9) and a whole part of the Convention is dedicated to the responsibilities of employers. In general, employers shall ensure that workers receive relevant information; chemicals are properly labelled and used with necessary precautions (article 11). Employers are also responsible to ensure that workers are not exposed to chemicals beyond exposure limits and if possible exposures should be assessed, monitored and recorded (Article 12). Certain provisions are also dedicated to transport of chemicals, disposal or operational control, duties of workers and exporting states (ILO 2003).

The impact of the Convention is limited because there are only 18 ratifying states as of 2015. Among them Brazil, Burkina Faso, China, Colombia, Dominican Republic, Norway, South Korea, Lebanon, Mexico, Syria, Tanzania and Zimbabwe and several EU states: Finland, Germany, Italy, Luxembourg, Poland and Sweden (ILO 2015). The membership of EU states is particularly surprising because the EU has specific norms covering rights and duties of workers. It might be stressed that the Convention is 25 years old and thus created before the EU single market which was almost finished by the time of Maastricht Treaty adoption in 1993. This may explain the attitude of Sweden which was that time not yet member of the EU, but not the later EU states which acceded to the convention during the last years: Italy in 2002, Poland in 2005, Germany in 2007 of Finland in 2014 (ILO 2015).

Next to the GHS system and the 1990 Chemicals Convention, there are some more conventions and instruments on the international level, which partly address the issue of hazardous substances. For example: The Convention for the Protection of the Marine Environment of the North-East Atlantic of 1992 (OSPAR Convention) deals with chemicals on its OSPAR List of Chemicals for Priority Action. In 2002 this list contained 42 persistent, toxic or bio accumulative substances of priority action. Among them was for example cadmium, mercury or phthalates including DEHP and DBP. The aim is to cease discharges, emissions and loses of these substances by the year 2002 (OSPAR Commission 2015). Due to EU countries and European Commission membership in this Convention, the system of selection of substances is based on REACH regulation and the prioritization on the Water Framework Directive.[3] The actions conducted under OSPAR are in line with Marine Strategy Framework Directive.[4] It is important to note that the protection of seas biodiversity is one of key aspects of the EU and since the adoption of the Lisbon treaty in 2009 the conservation of marine biological resources under common fisheries policy is under the exclusive competence of the EU (Article 3 TFEU).

Similarly to OSPAR, however on a lesser degree of complexity, there is The London Convention on the Prevention of Marine Pollution by Dumping of Waste and Other Matter of 1972 or more local The Convention for Protection of the Mediterranean Sea against Pollution of 1976 (known as the Barcelona Convention). There are also specific agreements regarding pollution from ships such as International Convention for the Prevention of Pollution from Ships of 1973 as modified by protocol in 1978 (known as MARPOL 73/78) which in its Annex II deals with noxious liquid substances in bulk or in Annex III with harmful substances carried by sea in packaged forms. As the sea environment is in danger from the oil related pollution in 1990 was signed International Convention on Oil Pollution Preparedness, Response and Co-operation (OPRC) which entered into force in 2005.

There are specific agreements and conventions regarding Air pollution as well, such as the Vienna Convention for the Protection of the Ozone Layer of 1885 and its Montreal Protocol on Substances that Deplete the Ozone Layer of 1989. However, they are out of the core of chemical management and not subject of this chapter. To sum up, there are many conventions and agreements signed by governments, touching various areas related to chemicals, its forms and properties. From the above mentioned "most visible" examples we can draw four important conclusions.

First, there is an existing platform on the global level to address potential threats posed by hazardous chemicals regardless the complexity and scope of the issue. Second, the geographic participation is more than obvious. While some countries especially the EU, OECD and partly Asia are signatories and ratifying parties, the voice of low income countries is missing (except for a few exceptions of rather

[3] Directive 2000/60/EC of the European Parliament and of the Council of 23 October 2000 establishing a framework for Community action in the field of water policy.

[4] Directive 2008/56/EC of the European Parliament and of the Council of 17 June 2008 establishing a framework for community action in the field of marine environmental policy.

regional agreements). Third, the EU is a very important actor with very complex attitude towards regulation of chemicals. This is due to the European commission participation on the conventions and the fact that most of the EU countries are signatories. Fourth, the regulation based on international agreements seems to be limited as low-income states simply may not have enough human resources and capacities to sign, ratify, implement and follow the development of the agreements and conventions which implementation is technically demanding and requires skilled personal. It is thus questionable whether the public sector in those countries may guarantee the safety from hazardous chemicals on its own. The lack of regulation is dangerous for public health and environment. On the other side, too much regulation is dangerous for companies and economic wealth of the state. This shifts us to the focus of private initiatives.

Some of the initiatives rose from the chemical industry itself. Probably the most known initiative is Responsible Care®, which represents the effort of chemical industry to improve environmental performance, health, enhance security and to communicate with stakeholders about chemical products. Among other commitments is to use resources efficiently and minimize waste, openly report on performance, achievements and shortcomings. Chemical companies and their associations should listen, engage and work with people to understand and address their concerns and expectations. Another aspect is cooperation with governments to develop effective regulations and responsible management of chemicals (CEFIC 2015a). The responsible care initiative has its origins in Canada where in 1985 the Canadian Chemical Producer's Association launched a campaign to improve public perception of the chemical industry and its safety standards. The core of this campaign was its critical approach to environmental performance and dedication to improvement (Topalovic 2007: 5).

Currently there are two important instruments. In 2006 the Responsible Care Global Charter was adopted addressing mainly safe use and handling of products along the value chain. The Charter presents Responsible Care core principles, fundamental features of national Responsible Care programmes and the approach in several issues including commitment to advancing sustainable development, improving and reporting performance, product stewardship or addressing stakeholder expectations about chemical industry activities and products (CEFIC 2006). In 2010 member federations adopted the European Responsible Care Security Code to describe practices to protect people, property, product, processes and information against any kind of criminal, malicious or cyber acts (CEFIC 2010). Since 2005 there are the European Responsible Care Awards awarded to companies with outstanding approaches to Responsible Care and Sustainable Development within eight categories: Communication, Distribution, Energy Efficiency, Environment, Occupational health & Safety, Process Safety, Product Stewardship and Security (CEFIC 2015b). The critical approach to its own activities and complex nature of Responsible Care together with commitments of the chemical industry created perception, that Responsible Care is considered as "self-regulation" of chemical industry (King and Lenox 2000: 699).

Above declared aims and efforts leading to higher safety and environmental standards are welcomed contributions from the chemical industry to modern chemical management. However, some critics note, that companies may use Responsible Care as a shield against and lobbying device against chemical regulation or that achievements of chemical remain superficial (Gibson 1999: 78). Moreover, as Aseem Prakash notes, implementation of Responsible Care was often problematic and without adequate control mechanism (Prakash 2000: 198). The evaluation of Responsible Care remains problematic as some of these arguments may be still valid and the level of control varies over countries. Responsible Care may also create the pressure on companies in the opposite direction. This is the example of the Czech company DEZA, which have been committed to fulfill principles of Responsible Care since 1994 (SCHP ČR 2014: 30). However, in 2008 there was an accident in the company and around 20 tons of naphthalene leaked into sewage. During the cleaning by hot steam it caused a strong odour and citizens of the city complained. This happened despite company management signing an agreement with the city where DEZA is located, on the mutual obligation to inform each other; a violation by DEZA (ČT24 2008) and against the principles of Responsible Care.

It seems that private "self regulation" has also its limits which lie on the interest of companies. We must keep in mind that companies are economic actors and as a such are interested in generating profit, keeping good image which helps them to do so and enhance good perception among its customers. Such behaviour may however result in practices which are unfriendly towards environment or public health and positive regulation must be accompanied by negative regulation "from above" which usually goes beyond the limits of voluntary standards and sets minimal standards of behaviour. It is an essential interest of every economic policy to regulate its subjects to avoid undesired externalities. That is why in the next part chemical regulation in the different parts of the World and the influence of EU chemical regulation will be discussed.

Part II
The Influence of REACH

Chapter 4
REACH Going to the EU Neighborhood

Second chapter explored the main norms regarding chemical regulation which are valid in the EU. Behind EU borders there are countries aspiring to EU membership. While some already have signed association agreements and opened enlargement negotiations with the EU, others are just potential candidates without early EU membership perspective due to political problems. This chapter deals with the EU neighborhood on its eastern borders, starting from the Western Balkans through Turkey and Moldova to Ukraine.

This group of the EU neighbours is very diverse, starting from the historical perspective, through geography and internal politics to economic structure. All these factors may influence the perception of the EU, its institutions, values and procedures. The management of chemicals is not an exception. For example, some countries in 2013 still had no register (inventory) for chemicals. Kosovo, Albania and Montenegro had no register but plans to develop it. On the other side Serbia had integrated register for chemicals, including biocides and plant protective products and Bosnia and Herzegovina had a register for chemicals including Safety Data Sheets and Macedonia had an inventory for licensed chemical companies (see ECRAN 2014). In other words, practices of chemical management varied significantly.

Despite great level of diversity, regarding to the response to REACH and other regulatory acts, we can divide those countries into two large groups. The first group consists of EU candidate countries, which have an obligation to adopt EU law in order to be able to join the EU. The second group consists of potential EU candidates which are not yet obliged to adopt EU laws, however they may do so if decided. From this perspective EU candidates should have imminent interest to implement EU chemical management law and thus their national legislation should comply to greater extent with EU rules compared to potential candidates. However, this political motivation is not the only one. From the economic perspective countries with EU oriented export of chemicals may adopt EU regulation faster than those with lower export to the EU. For this reason both political and economic dimensions will be assessed among these two groups of countries.

O. Filipec, *REACH Beyond Borders*, DOI 10.1007/978-3-319-54154-9_4

According to the above mentioned logic EU candidates where the influence of Europeanization is direct should have higher level of approximation in comparison to EU potential candidates. On the other side there are factors which might change this clear distinction. For example, the European Chemical Agency received 300,000 Euro in funding to prepare Balkan authorities for EU enlargement regarding chemical regulation. In total 107 officials from candidate and prospective candidate states took part in numerous seminars in order to strengthen their knowledge about REACH and CLP (ECHA 2012). Due to the IPA programme chemical legislation of those countries may be closer than differences in political and economic variables may suggest and minimalise the gap in the level of approximation between EU candidates and potential candidates. Assessing the impact of REACH in the Balkan region is of important value. As pointed out by Anastasakis, Europeization in Balkan is characterized as asymetric and EU driven process due to lack of national administrative capacities (Anastasakis 2005: 80). Is this also the case of REACH implementation?

4.1 EU Candidates

There are at the moment five EU candidate countries including Turkey (1999), Macedonia (2005), Montenegro (2010), Serbia (2012) and Albania (2014). Out of the five candidates negotiations which are structured into the chapters have already started with Montenegro (2012), Serbia (2014) and Turkey (2014). Out of these five only Turkey has opened negotiations within chapter 27 environment where chemicals belongs. In this sense it could be expected, that Turkey with the longest candidate status and opened negotiations will be much more advanced in REACH approximation than Montenegro or Serbia, despite certain set-back in EU-Turkish relationship after the July 2016 coup attempt.

There are other arguments supporting the expectation of advanced approximation. Turkey has very big chemical market compared to other candidates. Turkey's export value in chemicals in 2013 was $4916 million USD. Turkey exports around 25% of its chemicals to the EU-28. In this sense the EU is the biggest trading partner, exceeding Iraq (10%), Iran (8%) or Russia (4.5%). It is important to note that Turkish export is very diverse in its destinations and their relative share (World Bank 2013). It seems that the EU-Turkey customs Union plays an important role in REACH implementation because REACH may be considered as a trade obstacle as Turkish companies do not have the same rights as their EU counterparts. They are not for example allowed to submit registration dossiers to ECHA or they have to disclose some sensitive information by using "only representative" or registration (World Bank 2014: 152).

Preparations to adopt REACH in Turkey started under the IPA I programme (2007–2013) and following Technical Assistance Projects. Turkey assessed institutional and procedural structures which clarified the roles of institutions involved in chemical management. Existing registration and inventory system were enhanced by establishing a catalogue of chemical substances, substances in preparations and substances in articles. As the most important outcome of this period is legislation incorporating REACH (DEUT 2015). The Regulation of the Inventory and Control of Chemicals (CICR) entered into force on 26th December 2008, after consultations with the European Commission took place. The core of the regulation is considered notification of new and existing substances which are produced or imported in Turkey in quantities above 1 tonne per year (Bergeson et al. 2010: 1). The deadline has been set for 31 March 2011. However, as Turkey's experience with the EU regulatory system remained limited, Turkey decided to novelize this regulation with stronger regulation reflecting a higher extent of EU REACH. Areas for improvements were clearly visible in the information requirements and in the extent of regulation. For example under CICR only two tonnage categories are applicable. Producers or importers must provide only basic information on substances produced in a tonnage of up to 1000 tonnes per annum. Physiochemical, toxicological, ecotoxicological and other risk related data are required in a higher category over 1000 tonnes per annum. This system is very limited in new data gathering and data gaps may remain for substances produced or imported in lower volumes (Bergeson et al. 2010: 1). Moreover, the system is not yet fully operational.

Registration, however, is only part of the EU REACH spheres of influence. In this sense CICR reflects EU REACH registration requirements only to a limited extent and other spheres of influence (evaluation, restriction, and authorisation) are almost untouched by this regulation. In this way CICR is much more similar to pre-REACH Existing Substances Regulation (EEC) 793/93. In 2011 Turkey started the process of implementing EU REACH (REACH Centre 2015). In 2013 the draft of amending regulation called "KKDIK" (which means REACH written in Turkish[1]) was published proposing new deadlines for its registration. The new deadline is set for December 2018.

Despite this, CICR regulation is often considered as the "Turkish REACH" as it is accompanied by the Regulation on the Restrictions Relating to Production, Supply to the Market and Use of Certain Hazardous Materials, Products and Goods which entered into force together with CICR. This regulation reflects well the EU REACH restriction list, however it takes a much narrower approach with a limited impact. For this reason an early revision is considered in order to include more chemicals on the list (ChemSafetyPro 2015a). Much of the information requirements is covered by the Regulation on Preparation and Distribution of Safety Data Sheets regarding Dangerous Substances and Preparations which entered into force on 1 June 2016. This Turkish regulation fully reflects EU regulation 453/2010/EC. New Turkish classification and labelling legislation (known as the SEA) is less strict than CLP. For example by higher protection of confidential business information or longer periods for notifications which even allows the product to be placed for one month on the market without notification. On the other side periods for classification are much shorter than under CLP (Chemical Watch 2014).

---

[1] Registration (*Kaydı*), Evaluation (*Değerlendirilmesi*), Authorization (*İzni*) and Restriction (*Kısıtlanması*) of Chemicals (*Kimyasalların*).

However, full compliance with REACH and EU chemical legislation is soon unlikely because the perspective of early EU membership is small. Negotiations with Turkey are not satisfactory and Turkey faces significant problems including the occupation of Cyprus or meeting Copenhagen Criteria. Problems are not only on the side of Turkey as there is internal opposition against EU enlargement on Turkey within the EU which has increased after failed coup in summer 2016 and following democratic setback.

If Turkey with its $5 billion USD export is “the giant in the group”, there are two “dwarfs”. Albania and Montenegro have an export value of chemicals of only $12 million USD, respectively $11 million USD in the case of Montenegro. Both countries are strongly pro import economies in the chemical sector and it seems that all imported chemicals are used for domestic consumption as there is almost no chemical export from these countries. Montenegro exports its limited chemical products: 67% to Serbia and around 14% goes to Bosnia and Herzegovina. EU countries only share around 15% of Montenegrin chemical export (World Bank 2013). In Albania export of chemicals in 2013 was oriented mainly to Serbia (43%), Montenegro (16%) and Macedonia (7%), while EU countries covered by REACH represent around 22% of Albania’s export of chemicals. Moreover, the chemical trade is not diversified as there are only less than 20 export countries for chemical products (World Bank 2013). Economic data indicates that these countries may be reluctant to REACH approximation. However, there are political incentives.

Montenegro has an early prospect of EU membership and thus is encouraged in a political way. Montenegro has implemented REACH regulation to a large extent. It has designated the competent authorities and established the required infrastructure to ensure access by competent authorities to the databases of the ECHA. Montenegro established a system of inspections and enacted penalties in the case of non-compliance (European Commission 2013b: 14). Moreover, as of 2013 Montenegro launched a national helpdesk to assist manufacturers, importers and distributors on its responsibilities and obligations. It also identified the need to further build up the capacity to participate in the work of ECHA and its technical fora and expert groups. Montenegro also lacked the capacity to evaluate substances on the Community rolling action plan and is not able to submit dossiers for the identification of substances of very high concern. According to the Montenegro Screening Report form for 2013 (European Commission 2013b: 14), full implementation was expected in 2015. As such, Montenegro has successfully met this deadline, chemical regulation is not an obstacle in the terms of enlargement.

Albanian law on chemicals (Law no. 9108 of 17. 7. 2003) is only partially in line with REACH and because the control of hazardous substances is covered mainly by international conventions, further approximation is needed (IPHA 2012: 219). New legislation should be prepared and the office of chemical management should be established. Albanian progress is very limited due to the lack of personal capacities, the non-existent inventory of chemicals and the lacking methodology for the assessment of chemical properties (IPHA 2012: 219). Albanian law on chemical substances and preparations contains only 20 pages. This short document sets out the obligation of registration, however submitted data is basic. The current system is not aimed at data gathering; moreover it requires a proactive role from the Ministry of the Environment and the Council of Ministers which holds the burden of proof (see Article 13). On the other side, Albania may stipulate the obligation to register marketed substances (Article 9 and 10), evaluation of chemicals (Articles 7, 8 and 13) and classification (Article 5). The new law reflecting REACH is expected to pass in 2015.

In between 'these dwarfs and a giant' is Serbia with $768 million USD value of chemical export and Macedonia with approximately similar value 764 million USD. Serbia is exporting mainly to EU-28 countries (48% share) while export to Russia represents only 10%. Export to other Balkan countries is also considerable which will soon adopt REACH including Bosnia and Herzegovina (15%), Montenegro (10%) and Macedonia (8%).

In May 2009 the Serbian parliament adopted a new law regarding chemicals and biocidal products. In this piece of legislation Serbia adopted those provisions which allow Serbia to participate in the system without centralised EU procedures (ECHA 2011a). These include the incorporation of provisions regarding bans, restrictions, Safety Data Sheets, information flows in the supply chain and full transposition of the list of substances of very high concern. Serbia established its Chemical Agency in 2010 and help desk assisting companies to meet their obligations. The Serbian Chemical Agency started with a capacity building and transposing secondary regulation (ECHA 2011a). Serbia established its own electronic database which will help to identify products marketed in Serbia and products eligible for the registration under REACH. The idea is that once Serbia becomes an EU member, it would have no obstacles in joining the work of ECHA and fulfill its obligation (ECHA 2011a). Currently Serbia is very advanced in the alignment of their legislation on chemicals to the EU regulations with its law on chemicals covering the relevant parts of REACH.

Macedonia is the only country, where sales from the export of chemicals outnumbers import and sales is a very important source of state income. The chemical industry counts for 17.9% of its total export, which is far more than any other country in the region. Moreover, the Macedonian chemical industry is strongly dependent on the EU export as 88% goes to EU28 countries. The Macedonian industry is not diversified in terms of destination as 83% of its chemical production goes to Germany (World Bank 2013). These four combined factors makes the adaptation of REACH for Macedonia extremely economically important.

In the case of Macedonia (FYROM) the progress report of the European Commission is very general stating only that regarding the implementation of REACH "some progress was made" (European Commission 2013a: 53). However in one of the later reports the diplomatic language added that administrative capacity remains insufficient (European Commission 2014: 55).

The group of EU candidate states is also diverse group which differs in the significance of chemical industry, prospective of EU membership, but also administrative capacities and motivation. This results in different status of REACH adoption. It seems that Turkey with significant chemical industry has lost motivation to adopt REACH. The reason may lie in the no chance for early EU membership or the orientation of its export. On the other side Montenegro with not significant chemical export is very advanced in approximation and it can be expected that it will join EU as the first from the group of five. This is partly due to political problems within the other candidate countries. Serbia still has the problem of Kosovo, Albania has a long way to go to be part of the EU, Turkey has the problem of Cyprus and the Copenhagen criteria and Macedonia has disputes with its neighbours, especially Greece over its name. The slow progress of Macedonia in approximation is surprising. As a pro-export economy in the field of chemicals with a dominating orientation towards EU-28 it should have imminent interest to adopt REACH as soon as possible in order to improve status of domestic chemical industry.

4.2 EU Potential Candidates

Next to the EU candidates there are also countries, which are not yet granted candidate status and thus the negotiations is not open. They have no obligation to adopt EU acquis. Due to serious political problems, there is only theoretical chance that these states may join the EU in the next decade or a decade after. There is a huge political crisis in Ukraine affecting its territorial integrity and the potential future EU membership is limited by the Russian attitude and its policy towards the region. Russia is also an important variable in the foreign policy of Moldova which is still oriented to the east. Situation on the Balkan is no more positive. Complicated political and constitutional system of Bosnia and Herzegovina prevents effective approximation and thus early EU membership. The status of Kosovo is not as clear as part of the EU countries granted the independence of the new country and other countries still considers Kosovo a legitimate part of Serbia. The political incentives in those countries with the outlook for early EU membership are very week. However, the EU may play a normative role in the influence and this role might be supported by the economic incentives.

The chemical export of Ukraine is interesting. In 2013 the main export partner of Ukraine was Russia (30.5%) followed by Turkey (10%) and the USA (5.5%), while the EU28 counts for around 15% of chemical export sales (World Bank 2013). Due to ongoing conflict with Russia we can expect, that the share of Russia on chemical export of Ukraine will fall. As data from 2014 suggest, this does not happen as Russia counted for 34.9% of Ukraine export in chemicals in 2014 (World Bank 2014), so 4.4% higher than year before. At the first look this does not make sense. Better explanation could be find in absolute numbers. While in 2013 total export value of chemicals to Russia counted for $1315 billion USD, in 2014 it was $1055 billion USD (World Bank 2014, Jegelevicius 2013). So we can observe a drop by almost 20% between 2013 and 2014 while the total share has increased by 4.4%. This shifts the focus on total chemical export which decreased dramatically. While in 2013 total chemical export of Ukraine counted for $4.3 billion USD in 2013 it was around $3 billion USD. So in one year there was a dramatic decrease in the export of chemicals by almost 30%. The higher decrease in total export was mainly caused by falling export to Russia, though however it is falling slowly. This may be caused by the ownership structure of companies in Ukraine. So what are the implications for REACH?

Dramatic decrease in chemical export to Russia will create pressure on companies to search for the new markets. Due to geographic position of Ukraine the choice of Europe is likely. We can expect greater interest of Ukraine companies to export its chemicals to the EU 28 and thus greater approximation of companies to EU REACH requirements. As a result some companies have already voluntarily adopted the REACH standard. As noted by Vladimir Grishechkin, executive director of the Ukrainian Chemists Union (UCU), of the 30 companies belonging to the UCU, more than half comply with REACH provisions (Jegelevicius 2013). As estimated, Ukraine already implemented 70% of REACH provisions in its national legislation. Despite the early promise to adopt the remaining REACH provisions efforts have been frozen after former Ukrainian President Yanukovich halted the EU Association Agreement (Jegelevicius 2013). The example of Ukraine presents the situation where companies follow REACH without REACH being implemented on a national level. In this respect the price of REACH implementation on a national level is decreasing as the coercive logic of domestic companies is undermined by the early adoption on the side of companies and export reorientation to the EU market. This creates a favourable environment for future REACH implementation. Slightly different situation is in the Moldova.

Moldovan chemical export in 2013 for example, counted just for $163 million USD and was mainly oriented towards the Russian Federation (76%). Export to the EU countries is marginal (less than 10%). This makes Moldova the least economically important country in the terms of REACH adoption. However, even in Moldova there has been a political commitment to align chemical legislation with REACH since 2008 (MENR 2008: 352). Key objectives of sound chemical management to 2020 were identified in the National Program on Sound Management of Chemicals, including approximation with the EU chemical legislation. As Moldova is overwhelmingly an importer of chemicals, the Moldovan Ministry of Environment did not expect any difficulties in the adoption of REACH (Jegelevicius 2014). However, the progress in drafting national legislation implementing EU law is considerably slow.

Among other potential candidates is Bosnia and Herzegovina (BiH). Chemical export to BiH is slightly more significant than in Moldova, counting for $353 million USD in 2013. In the case of Bosnia and Herzegovina it is around 22% of the export that has an unspecified origin so data may not be relevant. However, export to EU countries represents 42%, to the Russian Federation 9% and to Montenegro 10% and Serbia 10% (World Bank 2013). In this sense REACH adaptation has also important economic aspect. However, the political line is problematic. The problem paradoxically rests in subsidiarity. Chemical regulation in Bosnia and Herzegovina is dealt with at the level of entities, not at the entire state level (ECHA 2011b). While the Republic of Serbia made considerable progress towards REACH adoption, in the Federation of Bosnia and Herzegovina old legislature from the Yugoslavia era is still in use. This created certain dichotomy in chemical management and harmonization on a central level is required.

The last potential candidate on the European continent is Kosovo. Due to its problematic status Kosovo is not part of the UN and thus is not a signatory of the Rotterdam, Stockholm or Basel conventions which is causing difficulties in the general extent of the issue. However, the Ministry of the Environment and Spatial Planning which is responsible for drafting chemical legislation received support from the European Commission under different programmes (TAIEX, IPA) and since 2014 has received support from ECHA. As the result of participation in various projects and networks Kosovo was able to partially harmonize its law regulations such as REACH, CLP, PIC, RBP and detergents with the EU law (Tahiri 2015). The example of Kosovo shows the importance of networking and external support for law approximation, making external support a considerable variable.

The perspective of the EU neighbours is not so negative as it seemed at the first sight. Out of five candidates, all countries are taking necessary measures in adaptation to REACH. In the case of Montenegro the chemical regulation approximation is supported by the early perspective of the EU membership as the country faces no serious political problem. The case of Serbia, Turkey or Macedonia is not so promising from the political point of view. However, economic dimensions support adoption of REACH. This is particularly the example of Macedonia, where export of chemicals is higher than import. Moreover, the vast majority of export goes to EU 28.

From the potential candidates Kosovo is most surprising. Thanks to networking and the generous support of ECHA Kosovo administration is able to comply with the EU chemical regulation to a large extent. The limits lie within the political dimension. This is also the example of Bosnia and Herzegovina and Ukraine. Bosnia and Herzegovina require substantial reform of its administration to enhance cohesion on the state level in order to comply with REACH. Ukraine and Moldova have to deal with Russia. In both cases approximation of laws may be supported by the changing orientation on the EU markets. Despite the various degrees of EU chemical regulation as reflected in the national legislation, REACH and subsequent regulation becomes reality in the EU neighborhood. However, in line with the conclusion of Othon Anastasakis (2005) also the implementation of REACH is rather slow and externally driven process. This thinking about REACH raises new question. How much is REACH reflected behind the EU neighbourhood by other EU partners?

Chapter 5
Europeanization of US Chemical Policy?

The EU represents for US companies a very important market in the terms of chemicals. In 2014 US chemical export counted for almost $164 billion USD while import in chemicals counted for $183 billion USD. In the terms of export, the most important partner according to percentage share are the remaining two NAFTA countries: Canada (14.62%) and Mexico (9.95%). On the third place is Belgium (7.51%), which is followed by Japan (6.08%) and China (5.9%). However, if we count EU28 as one subject, the share of the EU on the US export is 30.5% (WITS 2015).

The United States of America have been deeply involved in the first preparation of REACH and lobbied extensively for weak regulation since the preparation of White Paper on Chemicals. US activities and lobby strategies were aggressively opposing strong regulation of chemical industry. However, once REACH was adopted in its final form US companies started with adaptation to REACH and some US institutions developed cooperation with the European Commission and newly established ECHA. The main framework for cooperation is the Transatlantic Economic Council (TEC). There are many issues being solved at the TEC platform including information exchange about REACH and US chemical legislation, cooperation in OECD related activities, GHS harmonisation or exchange of personnel between ECHA and US Environmental Protection Agency (Berend 2009). Other activities include bilateral or trilateral meetings (together with Canada where issues are in many cases identical to those with US) or organization of High Level Regulatory Forum (Berend 2009). In November 2010 ECHA and US Environmental Protection Agency concluded Chemical Management Technical Cooperation Agreement aimed at substance information sharing and both actors with the European Commission signed Accord on development of chemicals policy (Simpson 2012).

O. Filipec, *REACH Beyond Borders*, DOI 10.1007/978-3-319-54154-9_5

5.1 Toxic Substances Control Act

Chemical regulation in the USA has been based on the *United States Toxic Substances Control Act* (TSCA) since 1976 which is operated by the Environmental Protection Agency (EPA). When enacting new regulation legislators faced a crucial question about the scope of the regulation. Should new legislation be retrospective and cover also old substances or should it be created only for new ones? The scope of regulation under TSCA presents a compromise, establishing distinct forms of regulation for "existing substances" and "new substances". The key element in this matter is the TSCA Inventory which collects basic information about substances manufactured or processed in the USA (TSCA: 218). According to the TSCAs definition *"a new chemical substance means any chemical substance which is not included on the chemical substance list compiled and published under section 8(b)"* (TSCA: 192). All substances which are not registered on the TSCA Inventory or are not subject to exemption are not allowed to be manufactured or imported. When TSCA entered into force in 1976 all substances existing at that time were subsequently grandfathered into the inventory. Thus the TSCA presumed that all substances marketed before December 1976 were *de-facto* safe in use.[1] This approach to the large extent undermined the distinction between both categories. Moreover, it did not improve on the lack of information about intrinsic properties of registered substances.

According to the TSCA the producer was obliged to send pre-manufacturing notification for new substances to the Environmental Protection Agency. When the EPA discovers some "unreasonable risk to human health or environment" the regulation took place. This could be managed in different forms: recommendations on labeling, recording and monitoring the use, limiting use or production. Total production or sale ban is also an available solution (TSCA § 6). By adopting regulatory decisions the EPA had to consider the effects of the regulated substance on health, magnitude of the exposure, exposure of the environment, benefits for various uses and the availability of substitutes and other economic consequences (TSCA, p. 211). In reality the EPA had only a limited time of 90 days to object to new chemicals after which the substance was placed on the TSCA Inventory and became an existing substance (Denison 2009a: 5). The limits were much more extensive due to a lack of information. The manufacturer's obligation to provide information about intrinsic properties relied mainly on voluntary measures and thus rarely generating new information (Applegate 2008: 732). The burden of proof was on the side of the EPA with a lack of information which soon led to a paradoxical situation. The EPA did not have enough information to prove unreasonable risks which lead to a regulatory underperformance.

The TSCA system did not establish effective motivation for data gathering. The EPA had authority to require testing but the testing requirement was subject to juridical review under which the EPA had to prove, that chemicals might present an unreasonable risk. As John Applegate points out, there are more than sixty-two

[1] It is estimated that existing substances counts for 99% by volume of chemicals in commerce (Applegate 2008: 732).

thousand chemical substances within the TSCA Inventory, but the EPA required testing of fewer than 200 chemicals (Applegate 2008: 738). Some of the tested chemicals were prioritized. Under section 4 of the TSCA a special committee was established for the creation of Priority list. On the list were substances considered for the promulgation of the rules. However, the committee had to consider relevant factors such as the quantities in which the substance or mixture is or will be manufactured, quantities in which the substance enters the environment, the number of individuals who are or will be exposed and the extent of exposure (TSCA, p. 197). The toxicity and other substance properties were not mentioned. The "information gap" was not the sole problem of the TSCA system. Michael Wilson and Megan Schwarzman (2009) identified two other gaps under the TSCA.

First, the "safety gap" caused by a lack of legal tools possessed by the regulator to identify potential health and environmental effects of hazardous chemicals and adopt appropriate regulatory measures (Wilson and Schwarzman 2009: 1202–1209). Due to a lack of appropriate regulation some states placed the priority on clear objectives for identifying and acted to control chemicals of concern. As a result the level of control differs within the USA (Denison 2009a: 2). For example California Proposition 65 of 1986 which passed by direct voter initiative prohibits businesses from knowingly exposing individuals to listed substances without proving a clear and reasonable warning (Proposition 65: 25249.6.).

Second, Wilson and Scharzman identify a "technology gap" presented by the lack of initiatives for green chemistry research, development and scientific advances (Wilson, Schwarzman 2009: 1202–1209). For example there was no regulation of nanomaterials under the TSCA which are gradually being placed on the market. Some studies already proved that exposure of some ultra-small particles could cause harmful effects on the human body including skin irritation; organ failures even changes in the DNA structure (see Sharma, Kumar and Dhawan 2012).

To sum up, TSCA presented shortcomings similar to EU countries, which were described in Sect. 2.1. Those issues include information gap, safety gap and technology gap. The gaps combined with the ill placed burden of proof and judicial review[2] made the system ineffective. These were realized by many actors, including both political parties which started to focus on reform proposals.

5.2 Reform Proposals

After 30 years of experience with TSCA and several amendments dealing individual substances[3] there was general consensus for crucial reform of chemical law. Some authors argue, that the REACH model for regulation was considered as a costly

[2] See for example the case Corrosion Proof Fittings v. EPA as summarized by Linda Stadler (1993) which led to failure of EPA to ban azbestos.

[3] For example 1986 Asbestos hazard Emergency Response Act, 1988 and 1990 Radon Acts, 2008 Mercury Export Ban Acts or 2010 Formaldehyde Standards for Composite Wood Products Act (see Schierow, Linda-Jo 2013).

alternative and anticipated that none of the "core elements" would be used in US chemical reform (Uyesato et al. 2013: 338). However, there were several proposals submitted to Congress which in some parts reflected REACH to various degrees. In 2010 there were two bills considered before both the House and Senate: The Toxic Chemicals Safety Act of 2010 (HR 5820) and Safe Chemicals Act of 2010 (S 3209), which was later revised and reintroduced as the Safe Chemicals Act (SCA) of 2011 (S 847). Despite considerable discussion none of these bills came to enactment due to other priorities in 2012 election year. Attempts to find the best solution for complex reform continued in 2013 when Democrat New Jersey Senator Frank Lautenberg and Republican Senator David Vitter from California introduced a bipartisan bill to modernize the TSCA. The so called Chemical Safety Improvement Act (CSIA) became the alternative to the Safe Chemical Act which were considered as more protective than the industry-backed CSIA (Shapiro 2013). However the Safe Chemical Act of 2011 also was updated and re-introduced as the Safe Chemicals Act (SCA) of 2013 (S. 696) by Democratic Senators Frank R. Lautenberg and Kirsten Billibrand. The next three sections are dedicated to further exploration of SCA, CSIA and finally also the 2016 Lautenberg Act in the context of EU chemical regulation in the areas of registration, evaluation and authorization of chemicals.

5.2.1 *Safe Chemicals Act(s)*

From the early beginning there were attempts to amend TSCA. At first, Senator Lautenberg introduced The Safe Chemicals Act (S. 847) which was later reintroduced as The Safe Chemicals Act (S. 696). At the same time Chemical Safety Improvement Act (S. 1009) was developed and also introduced to Congress (this act is explored in the following section). The distinction between SCA and CSIA is about the scope of reform. While SCA was proposing substantial change and new addition to TSCA, the CSIA would have amended TSCA significantly, however mainly without adding new provisions. Both suggested revisions were addressing the shortcomings of TSCA.

SCA would have established the obligation to manufacturer or process and to submit minimal data set regarding new chemicals or regarding existing chemical when required by the Environmental Protection Agency. These provisions were aimed at closing the data gap and gather enough information for further assessment. If the manufacturer of processor of chemical would have failed to deliver enough data, the EPA may have decided about necessary regulatory measures. According to Jerry H. Yen (2013) SCA would allow EPA to use testing methods effectively, quickly produce information and reduce animal testing (Yen 2013: 2).

SCA proposal in practice continued, as introduced by CSIA, to notify new use of chemical and adds new provisions requiring notification of any new use of chemical or notification in changes in production volumes (Yen 2013: 3). The EPA would be required in the 90 days period to categorize information based on submitted data and divide them into one of four categories:

- Substances of Very High Concern;
- Substances unlikely to meet safety standard;
- Substances with insufficient information; and
- Substances likely to meet the safety standard.

For the purposes of chemical assessment SCA would have deleted provisions regarding Interagency Testing Committee (ITC) which in TSCA system advised EPA about priority for testing chemical substances. ITC had to prepare a biannuall priority list of 50 chemicals which were given testing priority. SCA would have also directed EPA to establish its own system of prioritization of chemicals which will undergo testing. Despite ITC provisions are deleted, SCA builds on current TSCA provisions regarding the nature of substances. For these reason all substances currently categorized as "Substances to Undergo a Safety Standard Determination" would have been prioritized from 1 to 3 to risk assessment and already known chemicals would have been categorized as "Substances of very Low Concern" (Yen 2013: 5).

The restriction procedure is under SCA was to strengthen the agency. Production and use of chemicals would have been allowed only if the agency concludes that (based on risk assessment) the substance meets the safety standard and that there is reasonable certainty that exposure will not result to harm of human health or the environment. The EPA thus would have prohibit substances not to meet safety standard, substances of very high concern, substances assigned to a category "unlikely to meet the Safety standard" or substances lacking adequate information. In the case determination of EPA will be not followed, SCA gives to the agency tools to impose restrictions (Yen 2013: 6). This was a considerable shift in comparison to existing legislation. However, even in this case the review procedure before the court will matter, as decision of EPA whether a certain chemical meets safety standard or not may be subject of review.

However, despite safety determinations are subject of review, SCA stated that at all times each manufacturer and processor at all times bear the burden of proof in any legal proceeding. This was a considerable step forward in comparison to TSCA which was "crippled" by judicial review. Moreover, SCA went beyond and introduces new provisions. SCA would have required list localities where people are affected by exposure of chemicals and preparation of action planes to reduce exposure would have been required. Another proposed issue was to create a programme for supporting safer alternatives to dangerous chemicals and the replacement of hazardous substances. At least four research centers aimed at green chemistry promotion would have been established as well as children's environmental health research programme aimed at protection of children's health.

To sum up, SCA presented an important step in US chemical management which addressed many shortcomings of existing legislation. Compared to REACH, SCA has had data gathering potential based on risk assessment and allowed restriction of most hazardous chemicals and introduces more complex support of green chemistry. The burden of proof has been shifted closer towards manufacturers and operators, however there were some limits to its application which relies in review procedure and socio-economic analysis. However, SCA presented much more progressive alternative towards TSCA than second proposed reform act.

5.2.2 *Chemical Safety Improvement Act*

The Chemical Safety Improvement Act (S. 1009) was introduced to the 113th Congress. Regarding data gathering, CSIA was more similar to existing TSCA which does not require specific submission of data. CSIA has rather foreseen that the administrator (EPA) will develop the framework, policies and procedures for data collection and evaluation. After this framework would have been established the EPA would be entitled to demand data from manufacturers (Yen 2013: 2). Failure in data submission may have lead to penalties. Similarly to SCA, CSIA would have directed EPA to minimalise animal testing and encourage EPA to use non-laboratory testing methods (Yen 2013: 2).

Regarding the pre-manufacture notice CSIA was again closer to TSCA rather than SCA because it would have required notification only prior significant new use of chemical. Similarly to existing practice and SCA, even under CSIA EPA would have been required to categorize substances in three categories:

- Substances not likely to meet the safety standard;
- Substances needed additional information; and
- Substances likely to meet the safety standard under intended conditions of use.

The CSIA was preserving the current testing system based on ITC recommendations, however updates to the system increases the role of ITC. It would have also directed the EPA to establish the screening process of new chemicals which was aimed to determine testing priorities on risk based criteria. Based on these criteria, chemicals would have been evaluated as high or low priority for safety assessment and safety standard determination (Yen 2013: 5).

CSIA was in the area of restriction of chemicals closer to existing legislation than SCA as the producers would have been allowed to produce and commerce all chemical substances until the EPA determines that they did not meet the safety standard. The EPA would have been required to develop assessment criteria and rules for safety determinations. Restrictions were available to the EPA, however the EPA had to consider and publish the availability of technically and economically feasible alternatives, economic and social costs/benefits of alternatives or relative risks posed by those alternatives. In other words, the burden to act would have been on the EPA and effectiveness of restrictions might have been undermined by socio-economic analysis and proportionality of desired restrictive measures.

These measures may have been subject of a review procedure which was granted under CSIA to any person affected by CSIA actions. The burden of proof was to remain almost the same, however, judicial review would not been authorized for significant new use determinations, rules regarding Asbestos, lead-based paints or PCBs (Yen 2013: 8). This was a considerable move towards improvement of existing chemical regulation. However, in comparison to SCA the CSIA proposal presented a weaker approach towards chemical management and could have been considered rather a light adaptation of TSCA.

Both SCA and CSIA were far from EU REACH despite provisions is closer to REACH rather than TSCA. As mentioned in the beginning, REACH was considered as a very expensive alternative to TSCA and it is reasonable that US law makers have chosen revision based on existing regulation than revolutionary REACH. Despite this we cannot claim that REACH had no influence at all. Especially SCA addressed green chemistry and safer alternatives which might be adaptation to current trends REACH is in line with. Moreover, at a certain level of abstraction, even CSIA represents a shift towards REACH as the core of chemical management rests in information gathering, classification of substances based on risk assessment and restriction of hazardous substances.

5.2.3 LCSA

The main proponent of TSCA reform Frank R. Lautenberg died in June 2013 from viral pneumonia and reforms proposals soon became deadlocked. However, discussions and attempts to reform ineffective regulation continued and in 2015 republican senator John Shimkus and democrat Tom Udall introduced "Frank R. Lautenberg Chemical Safety for the 21st Century Act (LCSA)" which was on 22nd June 2016 signed by the president Barack Obama. Presidential signature finished almost 10 years long debate about creation of US sort of REACH. As presented above previous proposals reflected REACH to minimal degree and Lautenberg's Safety Act is no exception. There are, however, some signs similar to REACH.

First, the burden of proof has been shifted from the Environmental protection agency to producers and importers of chemicals which corrects main deficit of TSCA and shifts the paradigm in line with REACH (LCSA, section 4).

Second, before the chemical is allowed to enter market, EPA will be required to affirm the safety of new chemical. This is valid also for significant new use of existing chemicals (LCSA, section 5). Moreover, as pointed out by Richard Denison, EPA is required to consider and mitigate unreasonable risks of new chemicals under its conditions of use by including reasonable foreseen circumstances of production, processing, distribution use or disposal (Denison 2016).

Third, EPA has the authority to require additional information based on testing by producers in order to determine chemical risk (LCSA, section 4). Thus data gathering provisions are strengthened.

Fourth, similarly to REACH existing chemicals will be evaluated for their safety based on health risk, including most vulnerable groups such as children, pregnant woman or the elderly. Similarly to REACH SVHC substances, EPA is required to prioritize chemicals based on the safety risk (LCSA, section 6).

Fifth, similarly to restriction provisions of RECH EPA has the authority to phase out chemicals posing unreasonable risk for human health or environment. It has the authority to determine what chemicals poses unreasonable risk on new risk-based standards, leaving least burdensome cost-benefit safety standard (LCSA, section 6).

Sixth, there are green-chemistry provisions under new act as alternative testing methods and data sharing is emphasized in order to minimalise the need for tests on animals. LCSA involves provision aimed at reduction of testing on vertebrates (LCSA, section 4).

Seventh, new legislation sets clear and legally enforceable deadlines. Despite similar direction of new US regulation, compared to REACH it is still very weak version. For example EPA shall until 22 June 2017 establish criteria for prioritization of chemicals as high priority or low priority. By December 2016 EPA shall select and evaluate 10 chemicals from the 2014 TSCA Work Plan for Chemical Assessment and by December 2019 shall select and evaluate at least 20 high priority chemicals and at least 20 low priority chemicals while risk evaluation period shall take no more than 3.5 years (LCSA, section 6).

It is questionable how LCSA will work and how above mentioned principles will be set into the regulatory practice. For example EPA shall in the period of six months establish rulemaking on the conduct of risk evaluations which seems to be one of the key aspects of the new amendment (for recommendations of the American Chemistry Council see AMC 2016, p. 7). Similarly to SCA and CSIA the LCSA is compared to REACH rather conservative regulation taking into account specifics of previous domestic development. However, the indirect influence of "global standard" is evident even in the title of the act.

5.3 Other Acts

The chemical regulation in the US was insufficient as described earlier. For these reasons some states adopted their own regulation. Probably the best known is "Proposition 65" which refers to Safe Drinking Water and Toxic Enforcement Act of 1986, adopted by ballot initiative in November 1986. Proposition 65 requires the state to maintain a database of chemicals (currently contains around 850 chemicals) known to cause cancer or birth defects which are prohibited to be discharged into the environment. The list of substances is administered by the Office of Environmental Health Hazard Assessment (OEHHA), part of California Environmental Protection Agency (OEHHA 2013). Companies and persons doing business in the state of California are obliged to properly label hazardous chemical substances and follow instructions about discharges of these chemicals into drinking water, instructions on environmental exposures, occupational exposures or consumer products. The most visible is a warning written on tobacco products: "WARNING: Using this product will expose you to a chemical known by the State of California to cause cancer". The key instrument is a 60-day Notice of Violation which may be submitted by private party to the Attorney General Office which may review the allegations. Proved violation of Proposition 65 may be subject of considerably high fines. For example it allows to fine violation $2500 USD per exposure per day and average settlement costs approximately $65,000 USD. However some violations where more manufacturers are involved are fined much higher (for more information see State of California Department of Justice 2016).

The size of the Californian market and high fines for companies violating Proposition 65 lead to the so called "California Effect" where US companies in other states were unable to ignore California rules and have taken necessary measures to comply with the regulation (Wirth 2007: 97). Many lawyers benefited from Proposition 56 and related actions.

Chapter 6
REACH and the BRIC Countries

A very specific group of countries is formed by Brazil, Russia, India and China known collectively as the "BRIC" countries. All four countries are remarkable for their economic growth in the past two decades which made them important international actors. This is valid also for considerable growth in chemical production (especially in China) which accelerated the need for complex chemical regulation and participation on global instruments. However, chemical regulation in BRIC countries is experiencing significant problems (especially in India and Brazil while in lesser degree also in Russia and China) mirroring their domestic problems including lack of administrative, financial or technical capacities. For this reason, except initial interest in Russia, the response to REACH among BRIC countries was limited. This chapter explores chemical regulation in all four countries and where possible the changes into the context of REACH are placed.

6.1 Brazil

Brazil is a leading state in chemical production in South America and belongs to top 10 trading partners of the EU in the field of chemicals. The Brazilian chemical export in 2014 counted for $11.4 billion USD (WITS 2015). The top country for export was USA where Brazil exported 14.52% of its chemicals. Second most important country was Argentina (13%), which was followed by Canada (6.29%) and Paraguay (4.74%). The EU (EEA) counted for 20.7% of Brazilian export (WITS 2015).

The fast growing Brazil market is in deep contrast with regulatory measures which develops only slowly. As of 2015 Brazil has no regulation for industrial chemical management at the federal level, but only legislation covering some applications and specific uses of certain product categories including pesticides, disinfectants, food additives or cosmetics (Mākelä 2013: 22). However, the situation in some areas seems to be better, matching international standards.

O. Filipec, *REACH Beyond Borders*, DOI 10.1007/978-3-319-54154-9_6

Since 1978 there were guidelines on workplace safety markings and labelling under the Statute or Edict 3.214, NR-26. In May 2011 a new Statute 229 was published and Working Safety Marking Regulatory Standards were updated according to GHS Purple Book Criteria. According to this new statute, if there is no national harmonized classification table of hazardous chemicals, an international table can be used. Labelling standard mirrors UN GHS Purple Book Criteria. Safety Data Sheets must be provided for each chemical product classified as hazardous according to GHS (Chemical Watch 2011). Another field where chemical regulation is at least partly present is international transport.

Transportation of chemicals is covered by 1988 Decree 9611 on the Multimodal Transportation of Goods and Materials and the Decree 96,044 on Rail Transportation Regulation of Hazardous Products and Materials. Currently both acts are matching international norms (ICAO technical instructions, IATA dangerous goods regulations and IMDG codes). Some rules also covers workplace safety (Regulatory Norm 15 with its 15 annexes) touching restriction of benzene (annex 13A), hazardous and carcinogenic chemical substances (Annex 13) or annex 11 deals with occupational exposure limits for chemicals agents covering 205 chemicals (Chemical Watch 2011).

Very important aspect of harmonization on international level occurs due to the Mercosur integration and Brazil's active participation on international initiatives. Brazil is an active participant on various chemical fora and international treaties including the Stockholm Convention, UN Montreal Protocol and is active in the field of economic integration on the continent. Brazil plays a key role in Mercosur and is one of the core members (together with Argentina, Paraguay and Uruguay). Those countries (and some of Mercosur associates composed of Bolivia, Chile, Colombia, Ecuador, Venezuela and Peru) adopted common rules covering selected aspects of chemical management (Chemical Watch 2011). For example, hazardous materials are transported under the International Land Transportation Agreement of 1994 which is based on the UN recommendations. In 2007 they started to consider the adoption of GHS and after they added the additional module to GHS training for implementation of new GHS documentation. Mercosur chemical industry associations from Brazil, Argentina and Uruguay also formed the Chemical Industry Council (CIQUIM) which helps to improve competitiveness of chemical industry in the region (Chemical Watch 2011).

Thus, chemical regulation in Brazil could be considered rather sectoral than complex as industrial chemicals, their registration or assessment is not covered. For this reason the Ministry of Environment published in 2003 a document called "Brazilian Chemicals Management Profile" (BCMP) to support sound management of toxic chemicals. It aims to improve understanding of life cycle of chemicals and available solutions to chemical management. For this reason, the National Commission on Chemical Safety (SONASQ) was established to consider international standards. SONASQ is unique body which gathers representatives from the federal government, industry, NGOs and academia. One of the aims is to implement National Chemical Safety Program (PRONASQ) and give impulses for improvement in the field of chemical management. CONASQ cooperates with the Brazilian

Chemical Industry Association which deals with Responsible Care initiative (since 1998 is Responsible Care programme mandatory in Brazil and in 2003 was enhanced by independent verification process) and developed an Evaluation System for Safety Health, Environment and Quality to reduce risks related to the use of chemicals in all stages of life cycle. Also Pro-Química, emergency response service in the case of incidents similar to US Chemical Transportation Emergency Centre, was established. This service deals with incidents related to hazardous materials and dangerous goods (Chemical Watch 2011).

Despite the above mentioned measures progress is very slow. Compared to other BRIC countries, Brazil with its weak regulation is in similar position to India, dealing with very similar problems. Chemical industry in Brazil worry about the scope of new regulation, there is inadequate capacities and lack of experience. For this reason, it is not possible to enhance previous existing legislation (like in the US) but Brazil has to look at the foreign model of chemical management. REACH as well as in the case of Asian countries or the US seems to be costly alternative and for this reason Brazils consider to adopt Canadian inventory model and notification procedures. However, experts agree that adoption of any model in Brazil environment will require years for implementation (ChemAdvisor 2015). Things are changing slowly. It is expected that GHS will be applied in 2015 for substances and by 2023 also for mixtures (Mākelä 2013: 22).

REACH's inappropriateness for Brazil does not mean that Brazil cannot be inspired by REACH from the technical point of view. Some degree of classification or prioritization will be necessary as well as chemical assessment. It seems that it is early to talk about technical specifications of the future system, however the cooperation with ECHA may be enriching for Brazil as well as for trade between the EU and Brazil.

6.2 Russia

The EU represents Russia's biggest export market for chemicals. Out of $21.4 billion USD export in chemicals the EU counted for 29.5%. Regarding states, most of the Russian export went to Ukraine (11.6%), China (7.45%), Brazil (7.28%) and Finland which counted for 6.6% of Russian chemical export (WITS 2015).

Similar to the USA and China, the EU (DG Enterprise and Industry) established cooperation with the Russian Ministry of Industry and Energy under the Dialogue on Industrial and Enterprise Policy and the Roadmap of the Common Economic Space between the EU and Russia. The common issues cover information exchange on prepared legislation, recognition of data, support to comply with requirements or customs issues such as fees and length of procedures (Berend 2009). The EU has particularly supported Russia in changing its current system which demanded high responsibility of state authorities to the system where responsibility is placed on the industry (Berend 2009).

If the REACH was in the US context considered as very expensive alternative to TSCA (Uyesato et al. 2013: 338) in the context of Russia the adoption of REACH is impossible due to lack of resources and expertise. At the time EU adopted REACH, Russia still relied heavily on product standardization to ensure chemical safety. The system was based on Soviet era inherited product related specific mandatory standards called "GOSTs" or hygienic rules (SanPINs) and limited number of "technical regulations" (Teknicheskly reglamenty or TRs)[1] from the post-soviet era, which set up minimum regulatory and safety standards for a particular industry, specific sectors or group of products (Chemical Watch 2014). In other words, Russian chemical regulation was on the lower level and lacked single regulatory framework.

Russian system of chemical management was based on several legal acts:

- Federal Law of March 30, 1999 No. 52-FZ on sanitary and epidemiological welfare of population;
- Federal Law of July 21, 1997 No. 116-FZ on Industrial Safety of Human-Risk Industrial Facilities;
- Federal Law of January 10, 2002 No. 7-FZ on Environmental Protection;
- Federal Law of February 7, 1992 No. 2300-1 on Protection of Consumers Rights;
- Labor Code of the Russian Federation of December 30 2001 No 197-FZ; and
- Federal law of July 19, 1997 No. 109-FZ on Fisheries and preserving aquatic biological resources.

Federal Law on sanitary and epidemiological welfare of population and Government Decree about State Registration of Potentially Hazardous Chemical and Biological Substances of November 12, 1992 No 869 deals with registration of potentially hazardous substances, which is compulsory. This covers all potentially hazardous substances and biological compounds (including ingredients in the mixtures) produced or imported to Russia. As of 2010, the Russian register contained around 3400 substances, other 15,000 substances were investigated already before 1992 when register was established (Reihlen et al. 2010: 14). Substances are registered for certain period, usually 3 years. The register contains information about intrinsic properties of chemicals, its hazards for humans and environment and toxicity. Despite the register being public, access is limited to registered users who have to pay.

Registration starts with preparation of documents (dossier) including necessary information about intrinsic properties and toxicity. For this purpose companies are contracting accredited laboratories which are the only way for the substance assessment. After all necessary information is known (including classification of substance into hazard class or technical data regarding production such as production volume, purposes of the use, product specification or recommendations for handling) and registration procedure finished, relevant authority (Federal Service for Surveillance on Consumer Rights Protection and Human Well-being;

[1] In 2003 entered into force The Federal Act on Technical Regulation which replaced technical regulations by 1 July 2010.

RosPotrebNadzor) issues to the company certificate. In principle, the registration procedure is similar to EU registration under REACH. Registration is paid and fees vary according to character of the substance and time requirements on issuing certificate. There is no prioritization of substances for neither registration nor assessment under current legislation.

After the certificate is issued, the substances may be used on the market. New substances are under an observation period of 1 year when possible negative effects on human health and environment may be detected. All such cases are reported and quality standards updated accordingly. However, even high risk substances, which usually receives certificate only for the limited period of 2 years, may be used on the market despite competent authorities refused to release quality standards for safe handling of the substance (Reihlen et al. 2010: 17).

Federal law on Industrial Safety of High-Risk Industrial Facilities and relevant national standards (GOSTs) are the basis for classification and hazard communication. Separate GOSTs are dealing with classification of hazardous substances is various media, classification of dangerous goods, classification of hazardous substances in the manufacturing environment and other relevant areas such as pesticide or waste (Reihlen et al. 2010: 32). Since 2010 the labeling requirements reflects GHS classification principles under Federal Law "TR on Chemical Safety".

The Russian system of classification uses several classes for various dangers posed by substances. While the EU system is based on primary data from testing, the Russian system of classification is derived from the danger class which sets out requirements for handling the substance and possible restrictions (Reihlen et al. 2010: 33). For example Class I (extremely hazardous) and Class II (highly hazardous) chemicals may not be used in mixtures included for consumer protection while Class III (moderately hazardous) and Class IV chemicals (low hazard) may be freely available to customers. In GHOSt there are criteria on the substance characteristic and the class where substances fall (Reihlen et al. 2010: 33). Another type of classes is used for types of hazards. These includes flammable gases, oxidizing substances, flammable substances, explosive substances, toxic substances, highly toxic substances and substances dangerous for aquatic environment.

The Russian system knows also the Safety Data Sheets (Safety Passports) since the 1990s. However, for many years there was no legal obligation to use them and their use was on voluntary basis and their content and practice was also different from the EU standards (Reihlen et al. 2010: 39). In 2007 GOST 20222 was revised and since then Russian Safety Data Sheets shall be in accordance with GHS and REACH regulation (Reihlen et al. 2010: 39) and currently became a compulsory part or technical documentation of a chemical product. Safety data sheets must be distributed free of charge so all the customers have relevant data for safe handling of the substance (Reihlen et al. 2010: 39). Similarly to the EU, Safety data sheets in Russia began to play important role during registration process and are significant information tool.

In the case of restriction of chemicals (or possible pollutants) Russia has no specific regulation on restriction compared to the EU. As noted by Reihlen et al., the measures are not directly comparable, however they share some elements in the legislation (Reihlen et al. 2010: 42). Despite there is a basis to recognize extremely hazardous substances which discharge to environment is prohibited or they may be consider very dangerous to be placed on the market, the final decision is not formalized and depends mainly on the opinion of the assessor of the substance (Reihlen et al. 2010. 46). There are some specific rules for pesticides, which have separate registration system which works better then register of chemical substances. Authorities keep “List of pesticides and agriculture chemicals allowed for use on the territory of the Russian Federation”. Implicitly, all pesticides not listed are not allowed to be used. Russia also has specific framework for substances depleting ozone layer. Such substances are subject of licensing and some substances are prohibited under Russian law as a consequence of implementation of Vienna Convention and Montreal Protocol (Reihlen et al. 2010: 42).

The Russian system is similar in the area of registration to the EU, despite the lack of deadlines and prioritization of chemicals. Requirements on registration and the amount of information, including Safety data sheets reflects international standards and EU REACH principles. Assessment of chemicals in certified laboratories is also similar element with the EU system. However, the restriction of substances is of great concern as the Russian system fails to remove dangerous substances with very hazardous properties from the market. The system of classification and labeling of substances improved significantly. From rather chaotic labeling Russia adopted system implementing GHS which improved the communication of hazards to the consumers.

The newly adopted REACH presented a great source of inspiration for Russia as the drafters wanted to adopt as much as possible REACH provisions into newly elaborating Regulation on the safety of chemical products. The idea was that adoption of international standards will potentially in the future allow mutual recognition of registrations between the EU and Russia. For this reason Russian proposal was judged by EU experts and the REACH center was established helping Russian companies to comply with REACH (Kinnunen 2008: 38).

However, the idea that the EU will be open for mutual recognition was too optimistic. Moreover, full implementation of the complex REACH provision did not match Russian economic and administrative realities (Chemical Watch 2014). State obligations under the World Trade Organization required non-discrimination of domestic and foreign product which was considered as extremely burdensome for Russian companies oriented mainly on domestic consumption or former post-Soviet countries (Chemical Watch 2014). Costs to comply with REACH were higher than benefits associated with the access to the EU market. Due to lack of information and high demands of REACH some Russian companies considered to change their export destination. Igor Kukuškin, Executive Director of the Russian Chemists Union (RCU) said, that there were in 2008 plans to shift up to 20% of export activity to other countries (Kinnunen 2008: 39).

As a result Russia drafted proposals which reflect REACH only in a minimal way. In general, only provisions with direct product safety and which correspond with domestic economic and administrative capacities were adopted. Despite Russian reflection of REACH was limited, the regulatory draft included “no data, no market" provision, compulsory regulation, authorization of SVHC, chemical assessment and reports and the only representative. On the other side provisions related to data sharing and joint submission of dossiers were not introduced. Many Russian tools and processes were preserved including the list of chemical products, declarations of conformity and the system of accredited laboratories to certify the conformity (Chemical Watch 2014).

In 2009 the regulatory proposal was drafted with serious difficulties including lack of definitions, ambiguous concepts or imbalanced procedures. These include duplication of the control functions by different authorities or discretionary powers to grant or amend authorizations (Chemical Watch 2014). Ongoing difficulties have lead to further proposal wakening and further departure from REACH-like provisions. In this context REACH presented only soft source of inspiration for Russian drafters.

In 2013 the Board of the Eurasian Economic Commission considered safety regulation of paints and the regulation for safety of synthetic detergents and household chemicals. It also adopted draft decision on the safety of chemical products which already entered into force in February 15, 2015 and is thus considered an element of chemical regulation. This regulation established rules regarding classification of chemicals, inventory or substances and mixtures of the customs union, the notification requirements for the new chemical substances and mixtures, SDS and the rules for packaging and placing on the market (Wallerstedt and Vassileva 2013).

Russia improved its chemical regulation in the light of REACH. However the overall effect was limited due to domestic financial and administrative capacities. Due to ongoing economic integration in the region Russian norms and standards are becoming more important in the area of Eurasian Economic Union.

6.3 India

Indian export of chemicals in 2013 counted for $34.9 billion USD. Among top destinations for Indian export belongs the USA (18.75%), China (4.62%), Germany (3.91%) and Brazil (2.99%). The EU counts for approximately 22.4% (WITS 2015), which makes the EU the most important partner for India.

Indian legislation regarding management of chemicals was weak or non-existent and India was for a long time reluctant towards international cooperation in the field of chemical management. For this reason India lacked policy regarding registration of substances, national inventory of chemicals, law on restriction and banning of substances, detailed classification and labelling criteria and criteria for transport (Roberts 2012a). The only existing norms were Manufacture, Storage and Import of Hazardous Chemical Rules of 1989, which provide only a list of certain hazardous chemicals and partial regulatory measures regarding their use, or some other acts only partially addressing the issue of chemicals (Gas Cylinder Rules of 2004, Pharmacy Act of 1948, Poisons Act of 1919, Narcotic Drugs and Psychotropic Substances Act of 1985 or even Explosive Act of 1884). All together, 15 Acts and 19 rules (Fallström Mujkić 2012: 12).

Things started to change in India in the beginning of 2012 when the Ministry of Chemicals and Fertilizers called for drafting REACH-like regulation addressing all above mentioned shortcomings. For this reason the Indian Government prepared the Draft National Chemical Policy (Draft NCP-2012) which set up priorities in building new legislation. Among the main changes, a National Chemical Center (NCC) shall be established, which may have similar role to ECHA (Roberts 2012a). Next to the NCC there shall be Chemical Standard Development Organization to promote harmonization with international standards.

Drafting new legislation in India will be complicated due to a rather chaotic system of regulation and its fragmentation. First, it will be a real challenge to define competencies and connect new legislation with other existing acts. Second, Indian companies are skeptical about possible REACH-like regulation and so far the Indian position internationally. A similar attitude presented by the Indian government argued, that "one size fits all policy is not fair" and that developing countries should be treated in different way (Krishnan et al. 2013: 6). Because of localized and "unmatured" chemical market in India, REACH was considered as trade barrier for developing countries and international pressures a temporary problem. The Indian position towards REACH was "wait and see" (Raghavan, not dated). Third, because of fragmented legislation and low level existing standards, there is a long way towards comprehensive chemical regulation, which is almost being build on "green field" as there is no inventory of chemicals nor certified laboratory network established to conduct toxicological and eco-toxicological assessment (Raghavan, not dated).

Important aspect is the adoption of Restriction of Hazardous Substances legislation by India which will limit the amount of hazardous substances imported to India (Krishnan et al. 2013: 18). Since May 2014 there are six restricted substances (Lead, Mercury, Cadmium, Hexavalent Chromium, Polybrominated Biphenyls and Polybrominated Diphenyl Ethers) at the same concentrations as the EU does. However, compared to EU legislation there are several exceptions. For example this regulation does not apply to micro and small enterprises and the scope of products covered is limited (Krishnan et al. 2013: 22).

Regarding CLP, India was using since 1989 their own system based on Manufacture, Storage and Transport of Hazardous Chemical Rules which did not conform to GHS. However, it was expected to pass new legislation in 2012. For this purpose the Indian Government presented Hazardous Substances Classification, Packaging and Labelling Rules of 2011 which comply with UN system (Krishnan et al. 2013: 26).

India is good example of the country with fast growing chemical industry and improving chemical regulation which is facing many problems. For this reason Indian experience may be helpful for other countries which are building chemical regulation on the green field. In similar countries, REACH may present rational inspiration in various technical and procedural parts of the regulation.

6.4 China

Chinese chemical export in 2014 was worth $109 billion USD. Most of Chinese export went to the USA (11.72%) and India (9.68%). Japan counted for 6.78% and South Korea for 6.34%. EU is also among important trade partners counting for 15.6% (WITS 2015).

It is known that EU environmental regulation on several occasions influenced China which simply mirrored some norms. For example China solved its inadequate legislation on waste by mirroring EU legislation or adopted law on restriction of hazardous substances in electrical or electronic products (see Naiki 2010: 179). It is thus not surprising, that China closely observed (and sometimes lobbied) in the case of REACH.

After the adoption of REACH the EU has established cooperation with China under the Consultation Mechanism on Industrial Products and the WTO where employees from DG Enterprise meets personal of Chinese Standardization and Quality Control Body (AQSIQ). Some other projects were implemented by DG Enterprise, DG Europe AID and selected Member States and Chinese institutions including related agencies and ministries (Berend 2009). From the beginning there were many issues on both sides. China welcomed training on REACH to support Chinese export to comply with new requirements, gathered information about Guidance and IT and negotiations took place about recognition of test data from accredited test facilities. On the other side, the EU was interested in new notification of substances in China, GHS implementation in China or new law on toxic substances (Berend 2009). In support for compliance with REACH to Chinese Companies, China's Inspection and Quarantine Bureau established in 2007 the Chemical Inspection & Regulation Service (CIRS) which helps companies to do business between EU and China.

Chinese chemical regulation was for long time based on three acts:

- Hazardous Chemicals Safety Management legislation (1987);
- Regulation on Environmental Administration of the First Import of Chemicals and the Import & Export of Toxic Chemicals (1994); and
- Measures of Environmental Management of New chemical Substances (2003)

In January 2010 the Chinese Ministry of Environmental Protection released the Order No. 7, which updated Measures of Environmental Management of New Chemical Substances provisions and entered into force in October 2010 and since then is known as "REACH China".[2]

Regulation requires companies to submit notification about new chemical substance to the Chemical Registration Centre which is administered by Chinese Ministry of Environmental Protection. By the new chemical substance it is meant

[2] However, there are more accompanying laws including Decree 591 of the State Council of China in 2011; Regulations on The Safe Management of Chemicals in China; or various product specific measures such as Measure for administration of pesticide (2007); Measure for the administration of pharmaceutical precursor chemicals (2010) etc.

any other substances (including substances in preparations or articles and polymers) which were not listed on the Inventory for Existing Chemical Substances Produced or Imported in China (IECSC).

IECSC includes more than 45,000 substances, and stores both public and confidential information about them (around 3000 substances). Around 37,000 of chemicals has CAS number.

There are exemptions from notification for substances covered by other existing law, substances present in nature, substances of non-commercial purpose or unintentionally produced and special substances used in construction industry, pottery, glass industry or metal industry.

Notification is partly similar to REACH. Chinese companies may chose to submit new substance notification on their own or to hire local agent to notify the substance. Foreign companies have only one option in using "Chinese agent" which is similar as the Only Representative provision under REACH. By using this option a foreign company has obligation to submit appointment letter to the Chemical Registration Centre. Despite similarity with only representative under REACH, the conditions for entities in China to perform the role of Chinese agent are very strict (see Article 16). First of all, there is minimum registered capital of 3 mil. Yuan (approx. 350,000 Euro). Additional rules set up the condition that it should be company with address and fixed working place in China (no mailbox of fictive address is possible). Moreover, the entity has to pass annual industrial and commercial inspections, including inspections from the Ministry of Environment Protection. Agent must prove that personal has professional experience with regulation and that the company is capable to perform the agent task, have no officers in MEP and has no penalties for last 3 years.

Another similarity with REACH is in the rule "no data, no market". However in the case of China this rule is applied restrictively as only notification to CRC is necessary to ensure compliance. Thus "no notification, no market" is more appropriate expression. In the case that companies will not submit notification, they may be subject to warnings or sanctions, including fines, business restriction up to 3 years of business prohibition.

Notification is a key aspect of Chinese regulation. There are three types of notification based on the purpose of notification (Article 9). Under typical notification where the procedure may last between 8 months to 3 years (the higher production volume, the more data required), there are four notification levels based on production/import volume: first level (1–10 tons), second level (10–100 tons), third level (100–1000 tons) and fourth level up to 1000 tons (Article 11). Next to typical notification there is simplified notification procedure, which typically lasts approx. three months. This procedure covers all chemicals imported or produced in the volume under 1 ton per year, chemicals for scientific or technological research, polymers already listed in IECSC and other exceptions(Article 12 and 13). Both procedures require that CRC issues a registration certificate in order to start activities (Article 5). The fastest procedure lasting a few days is called the scientific research record which is used for notification of new substances used for research purposes produced/imported in quantities lower than 100 kg. Company submitting notification may start its activities immediately as no certificate from CRC is needed.

There are cases where serial and joint notification may be used. Serial notification may be used in the cases where there is similar molecular structure, similar use of the chemical and similar test data. Joint notification is allowed to be submitted by two or more notifiers and it is possible to issue the certificate for individual notifiers (Article 15).

Notification and certificate issuance is considered by the Evaluation Committee (see Chap. 3) and it is considered as the key aspect which plays important role for further chemical management and may represent further obligations for the applicant. The certificate will categorize the chemical as a new chemical substance or "hazardous new chemical substance" (which is almost the same as Substance of Very High Concern under REACH). Based on properties, the later category will be further classified as priority hazardous new chemical substances for environmental management. Requirement on registrant in post-notification period vary on the substance registered (see Chap. 4). While under "general new substances" registered the applicant is obliged to communicate with downstream users, submit activity reports, implement risk management measures, keep documents for 10 years, registrant of hazardous substances must comply with additional rules. These include the compliance with Measures for The Administration of Registration of Hazardous Chemicals or if the substance was classified as priority hazardous new chemical substance for environmental management there is obligation to submit annual plans, flow charts or reports on disposal information. Less follow up obligations apply to simplified notification or Scientific Research Record.

Once a substance is classified as hazardous chemical, then Regulation on Safe Management of hazardous Chemicals in China (2011) apply. In general, it covers the regulation of production, storage, use, sale, transport and marketing of the substance and requires the use of Chinese SDS (which is similar to GHS labelling system with noticeable differences. A system of permits is used to deal with hazardous chemicals (production permit, transportation permit, use permit, licenses for sales etc.). Currently there are around 4000 chemicals under Safe Management of Hazardous Chemicals legislation. In the case of non-compliance with SDS, failure of registration or operation without licenses strict penalties may apply up to hundreds of thousand RBM (Chap. 5). Even without penalties Chinese regulation is also costly. While the administration fee charged by CRC is around 200 RMB (approx. 30 USD), the testing fee may be several thousand RMB due to type of the tests and production/import volume (CIRS 2014a, b). This is another similar aspect to REACH.

Another similarity is labelling. Chinese GHS is strongly influenced by European standards. However, the Chinese system allows for the use of simplified labels, and have different norms for acute toxicity for gasses, pictograms may be black and red and there are 26 classes compared to 28 under EU CLP system.

Chapter 7
REACH and the Far East

Three of the four BRIC countries are located in Asia. However, there are more significant actors regarding chemicals industry. This chapter explores chemical regulation in Japan, South Korea, Taiwan and Singapore. The previous chapter demonstrated that lack of financial, administrative and human resources as well as technical obstacles with domestic regulation resulted in very weak response to REACH. Countries analyzed in this chapter have significant chemical industry and sufficient resources and expertise to overcome problems related to developing countries. What was the response of Japan, South Korea, Taiwan or Singapore to REACH and to what extent their domestic regulation is comparable to the EU's REACH?

7.1 Japan

Chemical export in Japan was in 2014 was worth $54,2 billion USD. Most important destinations for Japanese chemicals were China (25.22%), South Korea (17.85%), USA (14.9%) and Thailand (3.35%). The EU counted for 10.6% of Japanese chemical export (WITS 2015), which is considerably low.

Cooperation with Japan has been established under EU – Japan Regulatory Reform Dialogue. This dialogue was aimed at better understanding of REACH, its principles and enhancement of compliance of Japanese companies. The EU welcomed new information on Japanese legislative reform regarding chemicals and supported implementation of REACH elements (Berend 2009).

For a long time chemical regulation in Japan was based on The Act on the Evaluation of Chemical Substances and Regulation of Their Manufacture of 1973 called referred as „Kashinho“.[1] At the time Kashinho was enacted, it matched the highest level standards as it is considered to be the oldest chemical substances

[1] The Act on the Evaluation of Chemical Substances and Regulation of Their Manufacture (Kagakubushitu Shinsa Kisei Hou), Act No.117 of 16 October 1973.

O. Filipec, *REACH Beyond Borders*, DOI 10.1007/978-3-319-54154-9_7

control law in the world based on inventory system. Kashinho developed and was amended in 1986 and 1999. However, as pointed out by Yoshiko Naiki (2010) there were similar problems to those EU legislation has prior REACH: There was lack of information regarding chemicals, risk assessment was delayed and thus as of 2007 only 1500 out of 20,000 substances in the inventory have undergone risk assessment (Naiki 2010: 185).

These problems were addressed in the reform proposal which was created by the "Joint Committee to Review Chemical Substances Control Law" made of experts from the Ministry of Economy, Trade and Industry, the Ministry of Environment and the Ministry of Health, Labour and Welfare (Fallstrőm Mujkić 2012: 13). The final report suggested important changes in Japan's regulatory system: the construction of a system for risk assessment of chemicals based on their post-marketing status, an advanced pre-marketing evaluation system for the new chemicals based on the risk aspect and the handling of chemicals to be controlled by the strict risk control measures (Joint Committee 2008: 1). The Joint Committee created a complex reform proposal, establishing new system of chemical management (as presented on the Picture 7.1), which was implemented in 2009.

The reform proposal was soon adopted and entered into force under the Chemical Substances Control Law (CSCL) amendment from 20 May 2009 and was again amended later in 2011 in order to comply with international treaties. As already envisaged in the proposal, there are three responsible bodies for the implementation of the "new Kashinho": the Ministry of Economy, Trade and Industry; the Ministry of Environment and the Ministry of Health, Labour and Welfare. The new proposal covers industrial chemicals, as other categories are subject of different law (e.g. pharmaceuticals, cosmetics, fertilizers etc.).

The Japanese system is based on two main categories of chemicals. We can distinguish between new chemicals and "other chemicals". For this reason, new substances are defined as substances which are not classified as one of the following:

- **Existing Chemical Substances** (circa 30,000 substances) which were already on the market at the time when CSCL entered into force in 1973; or
- **General Chemicals** (Chap. 3) manufactured or imported in a volume more than 1 ton/year (circa 8000 substances); or
- **Exempted Chemicals** covering around 1500 substances; or
- **Class I Specified Chemical Substances** (30 substances including PCB, DDT, Aldrin or Endrin) which covers substances with persistent or/and highly bioaccumulative substances or/and substances with long-term toxicity for humans or predator animals. These substances require prior permission for manufacturing or import. In general, they are prohibited unless Cabinet order states otherwise and specific use is granted. In that case use is very restricted and must follow labelling requirements and other measures may be ordered (Article 17–34); or
- **Class II Specified Chemical Substances** (23 substances including Trichloroethylene, Organic tin or Carbon tetrachloride) which covers substances with long-term toxicity for humans, fauna or flora. There is obligatory reporting of planned import or use as well as reporting requirements for manufacture.

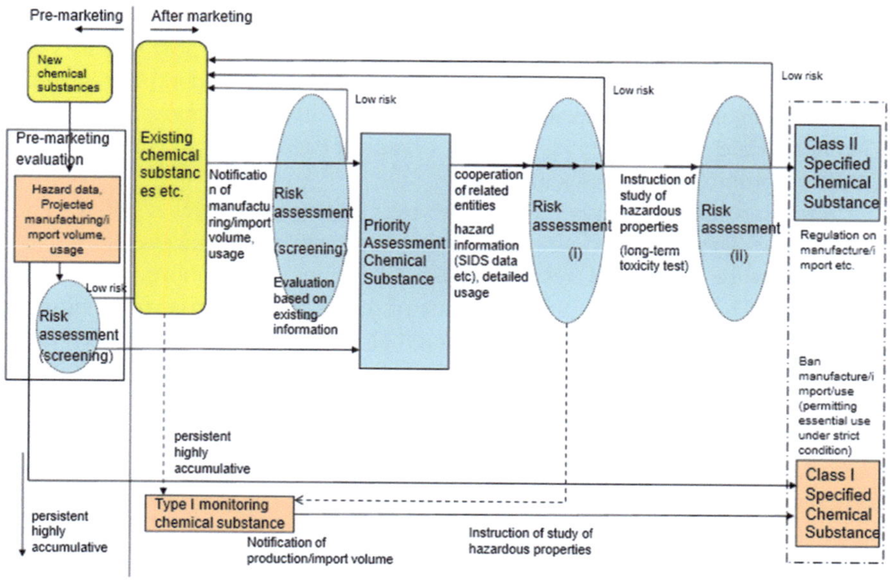

Picture 7.1 Revised system
Source: Ministry of Economy, Trade and Industry (2008)

Cabinet order may change the amount imported/produced, set specific guidelines and recommendations (Article 35–37); or

- **Priority Assessment Chemical Substances** covers substances suspected for long-term toxicity for humans, fauna or flora (Chap. 4). Usually after assessment is conducted they fall into the category of Class II Specified Chemical Substances.

Notification of new substances[2] is key aspect of the system (Article 3). Types of notification vary depending on the substance (polymers and non-polymers), volume (small volume up to 1 ton/year; low volume up to 10 tons/year) or concern. For this reasons we have several categories: Full notification for polymers and Full notification for non-polymers produced in quantities more than 1 ton per year, Small volume exemptions, Low volume exemptions, Exemptions for polymer of low concern, Exemption of substances specified by Cabinet order and Intermediates for export only. In some cases (Small volume exceptions or Low volume exceptions) there are only minimal data requirements, however usually prior verification and exemption certificates are required.

[2]The new chemical substance should be also notified according to Industrial Safety and Health Law (ISHL), which entered into force in 1972 aimed the protection of workers from chemicals. It regulates the use of „new substances "as well as other „prohibited", „permitted", „indicated "and notified substances used on the workplace and established ISHL inventory which is in some aspect parallel to inventory under CSCL.

Based on data provided by companies[3] during the notification procedure and based on the evaluation of government, substances produce in low volumes and have not hazardous properties are confirmed and allowed for import or production. The Government may also decide, that the chemical should be removed to Existing Chemicals Category, Priority Assessment Chemical Category or directly to Class I Category (prohibit the substance).

Under Existing Chemicals Category are over 30,000 substances which are considered safe. There are also new chemicals after evaluation. However, these substances may be subject of review by the Government and moved to Priority Assessment Chemicals Category (in the case of suspicion of low risk) or Monitoring Chemicals Category (in the case of suspicion of high risk or unknown risk). Once chemical is under priority assessment (Article 9–12), the manufacturer or importer of the substance has obligation to deliver another information regarding production quantities, usage of the substance or shipment details. If there is long-term toxicity for humans, fauna or flora, then substance is placed among other Class II Specified Chemicals. It this sense Priority Assessment Chemicals are "candidates" for Class II chemicals category.

Chemicals in the Monitoring Chemicals Category (37 substances including cyclodecane) are assessed for long-term toxicity for humans or predator animals (Article 13–16). If some persistent or bio cumulative properties are discovered, then the chemical is placed among Class I Specified Chemical Substances. It is worth to note that Japanese assessment is targeted at toxicity, bioaccumulation and persistency. Obligation for importers and manufacturers are similar to SVHC under REACH provisions (CIRS 2014b).

Both Priority Assessment Chemicals and Chemicals in monitoring deals with chemicals having dangerous properties or suspected dangerous properties. Next to the notification the prioritization of chemicals is another key aspect of Japanese system similar to REACH (Naiki 2010: 185). Picture 7.2 shows how Japanese Chemical Substances Control Law works.

Similarly to other laws Japanese CSCL under Article 19 sets sanctions. Permission to manufacture or import chemicals will be not granted to person who has been sentenced to a fine or severe punishment for violation of CSCL in the last 2 years or even a juridical person where the officers carry operations related to CSCL. Also GHS provisions are implemented in Japan under Industrial Safety and Health Law which requires classification and labelling since 2006.

In summary, Japanese CSCL is only in few minor aspects similar to REACH in the area of "prioritization" of chemicals where priority assessment is given to substances with proven hazardous or unknown properties. Both systems are aimed at data gathering, however under REACH the burden of proof including chemical assessment is on the chemical industry. While there are some information obligations

[3] As noted by Naiki (2010) industry responsibility is not explicitly stated in contrast to REACH and the burden on industry is much smaller than in the case of REACH. For example while in the EU industry is required to submit Chemical Safety Report, in Japan chemical safety assessment is conducted by the government (Naiki 2010: 186). The nature of the system thus differs in this way.

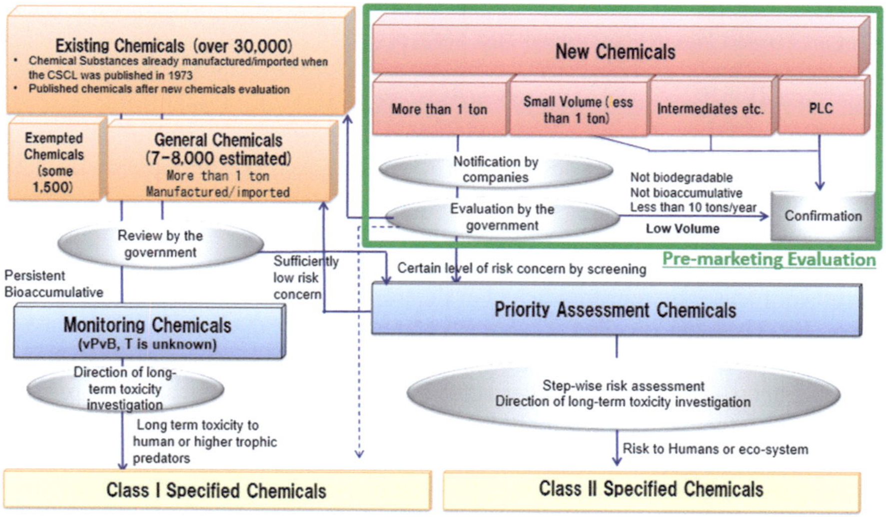

Picture 7.2 Chemical Substances Control Law System
Source: Ministry of Economy, Trade and Industry (2013)

under CSCL there is also part of responsibility on the government to generate information. However, the "identify and rule" system based on Cabinet orders seems to be more effective than the US system where burden of proof was on EPA with limited tools. System is thus much more similar to Canadian model than REACH system (see Sect. 9.1).

There are several reasons for it, as presented by Yoshiko Naiki (2010). First, REACH was not very compatible with Japanese domestic structure. Second, there was lack of industry support for the EU model, despite the fact that Japanese companies have to comply with REACH in order to continue export to EU. Third, it was not necessary to adopt comprehensive regulation because important information were being disclosed in the EU under REACH and fourth, there was a weak response by NGOs in support of REACH-like regulation (Naiki 2010: 190–193). It seems that the pragmatic response favoring existing experience with Kashinho lead to partial reform which was in some aspects are inspired by REACH but left the system still similar to the Canadian approach in chemical management, which will be revealed in Chap. 9.

7.2 South Korea

Chemical export for the Republic of Korea in 2013 was worth $38.4 billion USD. Korea exported its chemicals mainly to China (48.2%), Japan (7.1%), USA (4.51%) and India (4%). The EU export counted for 4.3% (WITS 2015), which is less than in the case of Japan.

For a long time the most important document in South Korea was the South Korean Toxic Chemicals Control Act (TCCA). TCCA is one generation younger than the Japanese Kashinho. TCCA was adopted in 1991 and several times amended (the last revision was by the Act No. 895 from March 2008). Under the act responsibilities are divided between the Ministry of Environment where its Korean Chemicals Management Association (KCMA) is responsible for industry declarations and issuing certificates and for National Institute of Environmental Research (NIER) which is responsible for new chemicals notification.

However, since January 1st 2015, TCCA is divided into two documents. The first document focuses on registration and evaluation of substances and is called "Korea REACH". The other is called "Chemicals Control Act" (CCA) which is specialized on the control of hazardous chemicals. Next to the TCCA (respectively Korea REACH and CCA), there are other acts covering specific chemicals. For example medicine product are covered by Pharmaceutical Affairs Act, cosmetics falls under Cosmetics act, fertilizers falls under Fertilizer Control Act, special acts are covering high-pressure gasses, firearms, swords and explosives, livestock and fish feed, agrochemicals, drugs etc. (see Article 3 of TCCA). It has been estimated that the new law will affect around two thousand substances and three thousand companies. Some of them, already made steps in compliance with the EU REACH (Fallstrőm Mujkić 2012: 12).

In this chapter TCCA is first explored which is necessary for understanding the development of chemical management and then are investigated the most important aspects of Korea REACH and Chemicals Control Act.

Under TCCA there are four kind of chemicals regulated (see Article 2 for further specification):

- New Chemicals, which shall be evaluated and notified before import or manufacture if the volume is more than 100 Kg per year/producer (these substances are considered as "notified substances" under Korea REACH;
- Toxic chemicals, where report requirements are higher and include usage, import or production quantities etc. These chemicals requires due labelling and packaging;
- Substances under observation where kinds, usage and annual quantities of manufacture or import shall be reported to KCMA if annual import exceeds 100 kg or the substance is not for research and development purposes; and
- Restricted or Prohibited Chemicals where permission is necessary for import, storage, use or export by KCMA as annual import exceeds 100 kg and the substance is not for research and development.

As mentioned earlier the Korean system distinguishes between "new" and existing "old" chemicals, which were marketed (domestically commercialized) prior February 1991 under the Ministry of Environment. For this reason, there was established the Korea Existing Chemicals Inventory (KECI), which currently maintains information about 42,000 chemical substances. The following picture shows how the notification system for old and new substances works.

Similarly to Japan, notification and data submission are key aspect of the regulatory system. The scope of data submitted varies according to procedure. There is a simplified notification for substances induced into KECI before 1991 (requires only acute toxicity and genotoxicity tests), semi-full notification for new chemical substance with volume below 1 ton per year (to the previous tests adds also in vitro chromosomal test and biodegradability test) and full notification for all other substances (previous additions were Acute fish, daphnid, algae toxicity, skin and eye irritation and skin sensitization). The basic requirements for all types of notified chemicals are also data regarding environmental release and physic-chemical properties (see Chap. 2) (Picture 7.3).

There are several exemptions from notification of new substances, which includes substances produced/imported in the quantity less then 100 kg/year; chemicals which are considered as inseparable part of imported machine; site limited research and development substances, chemicals in KECI marketed before 1991, block copolymers, graft polymers, polymers of low concern, new chemicals for exports purposes, chemicals to be used as non-isolated intermediaries or new chemicals used in closed systems. Excepted new chemicals require that KCMA issues Certificate for Exemption. Other new chemical substances, which are not exempt from notification requires confirmation by NIER (CIRS 2015a, b).

Korean REACH[4] was first published in May 2013 and entered into force in January 2015.[5] It deals with registration of chemical substances, screening of hazardous substances, hazard and risk assessment of products containing chemical substances and hazardous substances and information sharing. In some aspects, K-REACH is similar to TCCA. First, it does not apply to same scope of specific chemicals e.g. pharmaceuticals; narcotics, agrochemicals, fertilizers, explosives, feeds etc. (see Article 3). Another similarity is the definition of "new substance" which was not placed on Korean market before 2 February 1991 and notified by the Ministry of Environment on December 23, 1996; and chemical substances which have undergone examination of toxicity after February 2 1991.

The new substances shall undergo registration procedure (Article 10), which is many aspects (data submission) similar to TCCA and manufacturers/importers has to submit Annual report and report quantities and uses of the substances. This obligation covers also importers and manufacturers of existing chemical substances which exceeded 1 ton per annum (respectively they are imported in lower quantities but cause damage to human health or the environment). Exceptions from submitting annual report or registration (Article 11) is similar to those under TCCA exemptions from notification of new substances although the scope is limited (research and development purposes, substances contained in a product without function being released, inseparate part of the machine etc.). A person who fails to register a substance is not allowed to import or manufacture the substance and the Minister of

[4] In full The Act on the Registration and Evaluation of Chemicals (AREC).

[5] Preparation for K-REACH implementation consisted of public consultations and several preparatory steps (see CIRS 2014a, b).

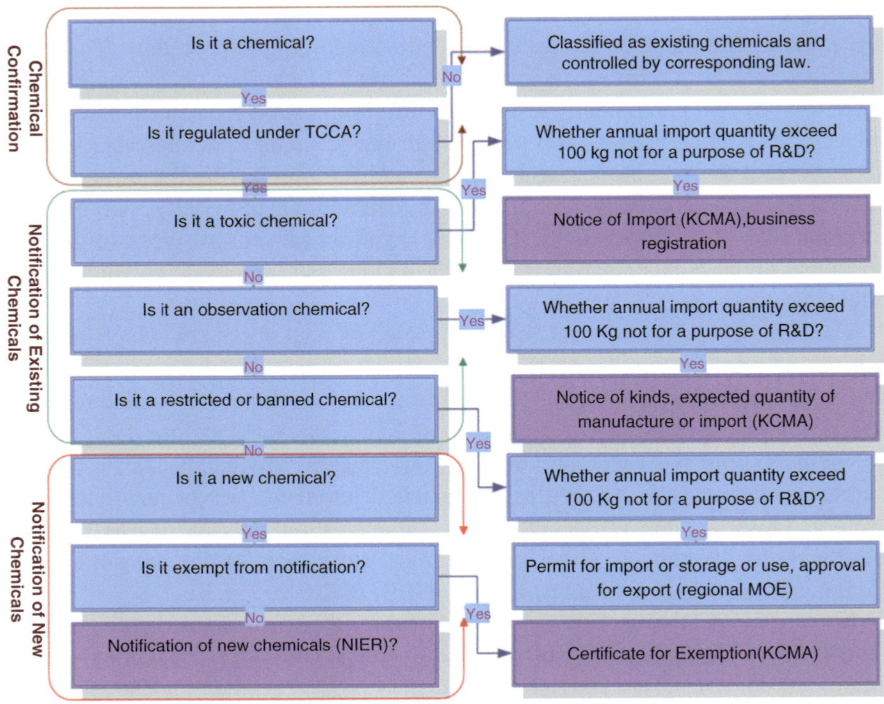

Picture 7.3 TCCA notification system
Source: Chemical Inspection & Regulation Service (2015)

Environment may order to suspend or cancel the process related to chemical (Article 13). This provision is similar to "no data, no market" procedure under REACH.

It is important to note that annual reporting covers also polymers. The scope of data submitted during registration varies due to production volume and type of the substance. Specific requirements may be introduced by Ministerial or Presidential decrees. For example, substances which exceed 10 tons production/import volume requires more risk associated data including exposure scenarios. However, required data may be reduced according to decrees (new substances produced/imported under 1 ton/year, polymers of low concern etc.).

Similarly to EU REACH, a foreign company may appoint the Only Representative (qualification is decided by the Ministry of Environment) which will fulfill the duties regarding reporting and may use joint submission of registration (Article 16). Another similarity with REACH is the deadlines for registration set to 2 years from publication date; 5 years from publication date; and 8 years from publication date. There are also separate deadlines for conducting risk assessment which is compulsory for all substances produced/imported in quantity 10 tons or more and for substances deemed as requiring risk assessment.

Assessment deadlines are based on tonnage according to Article 24. While chemicals produced in tonnage higher than 100 tons per year has deadline in 1

January 2015, substances produced in a quantity between 70 and 100 tons per year has a deadline on the 1st of January 2017; substances produced in a volume between 50 and 70 tons on January 2018; substances produced/imported in quantity between 20 and 50 tons per year have a scheduled deadline of January 2019 and the lowest tonnage substances between 10 to 20 tons per year have a risk assessment report submission for January 2020.

Based on the results of assessment, Ministry of Environment may classify the substance as hazardous, and specify as (according to articles 25–28):

- **Toxic Substance** (intention to import should be notified including type of substance and its usage);
- **Authorization Substance** (potentially hazardous, requiring permission before manufacture, import or use);
- **Restricted Substance** (highly hazardous, generally banned but with permission used for specific purpose);
- **Prohibited Substance** (highly hazardous, not allowed in general, however when used for research and development permission is required).

Following Picture 7.4 shows the process of registration and evaluation under K-REACH.

Specific provisions covers transport and related supply chain communication and product management, including risk assessment (article 33) of consumer products and biocidal products.

Complementary to K-REACH is Chemical Control Act (Law No. 11862) which entered into force at the same time as K-REACH.[6] It focus on the control of all phases of life cycle of chemicals, including prohibited substances, restricted substances, substances subject to authorisation, toxic substances and accident precaution substances (new chemical substances are the subject of K-REACH). Manufacturers and importers under CCA shall submit a "Written Confirmation of Details for Chemical Product" to KCMA prior to manufacture or import if one of above mentioned types of chemical will be produced or imported (Article 9).

The Act covers safety management of hazardous chemical substances including hazardous chemical handling criteria (Article 13–22) and installation and operation of hazardous chemical handling facilities (Article 23–26), business operators of hazardous chemical substances and preparation for and countermeasures against chemical accidents.

Classification and labelling obligations in Korea fully match GHS standard under Classification & Labelling of Toxic Chemicals (Public Notice 2008–26). Substances need to be classified and properly labelled since July 2011 and mixtures since July 2013. The system of classification and labelling has been slightly changed in 2012, improving information gathering related to Safety Data Sheets and to be more consistent with EU CLP regulation (CIRS 2015a, b).

---

[6] On January 2015 CCA was partially revised and Publisher as Law No. 13035, implemented on 21 July 2015.

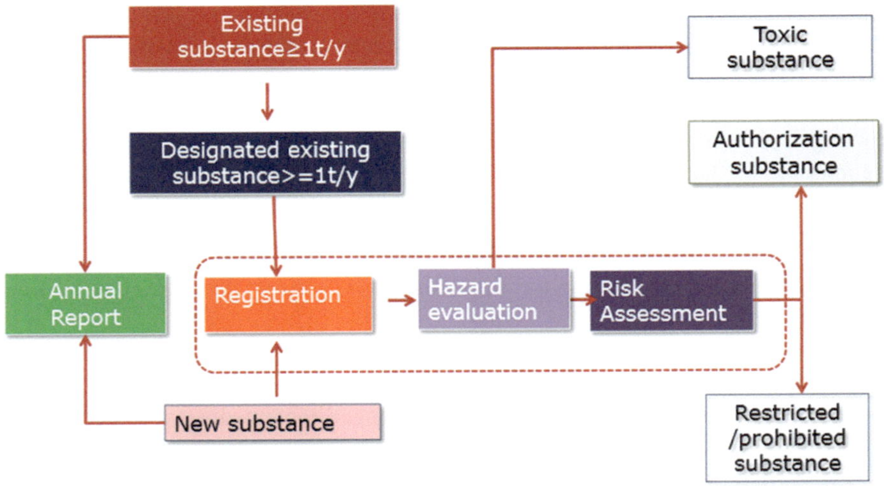

Picture 7.4 Registration and Evaluation under K-REACH
Source: Chemical Inspection & Regulation Service (2013)

Penalties apply to various violations of K-REACH. For example a person who sells product which has been banned due to hazardous properties may be sentenced to 7 years and fine of 200 million won (circa 170,000 USD). False registration or import without registration may lead to a 5 year sentence in prison and a 100 mil. Won fine (see Article 49–53).

Despite K-REACH in many aspects adopting the same direction as EU REACH, there are noticeable differences which were published on the Chem Safety Pro website. First, in the original document or EU REACH registration of new substances produced in small volumes under 1 ton/year is not required while under K-REACH those chemicals fall from the beginning under simplified registration. Second, even in the complexity EU REACH goes higher as it requires registration of all chemical substances, while K-REACH only new substances and designated existing substances. On the other hand, submission of Annual report is not required under REACH so Korean authorities have yearly updated data. Third, both legislation differs also in its approach towards polymers. REACH requires registration of monomer or reactant, while K-REACH requires registration of polymer itself. Fourth, many issues which are dealt in specific way under EU REACH are subject of application of confirmation of registration exemption procedure. For example, substances for research and development purposes are in the EU subject of PPORD notification.[7] This procedure under K-REACH covers also non-isolated intermediates and surface treated substances, which are fully exempt from EU REACH.

The procedures apply also to export-only substances while under EU REACH they are treated as regular substances. Fifth, biocides are not covered under EU

[7] Product and Process Oriented Research and Development exemption from compulsory registration under REACH can be prepared in IUCLID tool.

REACH as they are part of separate legislation while K-REACH covers them and has even higher data requirements for biocides than for regular chemical substances. Six, chemical consumer products are more developed under K-REACH (notification, safety and labelling standards for product or risk concern) while EU REACH deals mainly with registration of ingredients (ChemSafetyPro 2015a).

Due to differences K-REACH represents the most similar approach to chemical management to EU REACH. Some aspects of K-REACH address issues which are not originally covered by EU REACH, but REACH following subsequent legislation covers also issues previously not addressed by EU REACH (low volume substances, biocides, polymers etc.). Similarly to China and Japan also South Korea in the past has used some EU norms to improve national environmental regulation as the case of restriction of hazardous chemicals in electronic products shows (Naiki 2010: 182), however the way how Korea shifted its chemical management to EU REACH is unprecedented.

7.3 Taiwan

Taiwan imported in 2014 to the EU chemicals worth of 1.2 billion EUR (DG Trade 2015b: 2). If we consider that overall export from Taiwan in 2013 was around $300 billion USD and 7% were chemicals, it means that 1.2 billion EUR EU import counts approx for 6% of chemical export from Taiwan.[8]

The regulation of chemicals in Taiwan is covered by the Toxic Chemical Substances Control Act (TCSCA) which was adopted in 1986. Since then TCSCA has been several times amended, most comprehensively in the last amendment by Presidential Order which entered into force on 11 December 2014. According to Chemical Inspection & Regulatory Service Taiwan revised its legislation due to pressure caused by the fact, that other actors in the region (Korea, China and Japan) already passed new law aimed at new substances and there was real danger that Taiwan will became testing ground for new substances. The pressure was highlighted by the food plasticizer contamination incident in 2011 (CIRS 2014c).

The system is based on distinctions between new substances and existing substances. New chemical substances are those which are not listed in the national existing substances inventory (ECSI). Producer or importer has legal obligation to register such substances 90 days prior production or import if the criteria of exception are not met (article 7–1).

Those cover naturally occurring substances, substances accompanied in the machines, inseparable intermediates, chemicals for national defence purpose, chemical substances under customs supervision, waste, incidental reaction products, mixtures, articles, polymer applicable to the 2% rule listed in Taiwan ECSI and

[8]There are no data provided under WITS so it was necessary to use alternative method based on combination of data which are available at Index Mundi 2014 and Trading Economics 2015. Due to various resources used calculation is only approximate.

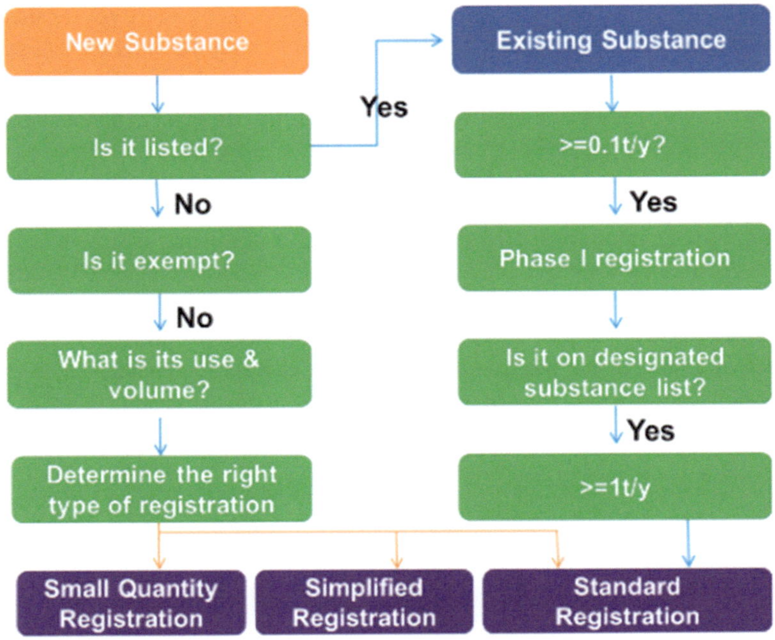

Picture 7.5 Registration process
Source: ChemSafetyPro (2015c)

other already regulated chemicals by existing law. After 5 years new chemical substances are added into ECSI. Only designated existing substances on the list prepared by authorities are subject of registration. Picture 7.5 shows the process of registration.

All existing substances imported or produced in a quantity higher than 100 kg are subject of Phase 1 registration where importer/producer is submitting basic information about the chemicals (article 7–1). Based on this elementary information some chemicals are due to its quantity or hazard placed on priority list. Those substances undergo Phase 2 registration which is called "standard registration". Based on substance type and production/import volume there are three types of registration: standard registration, simplified registration and small volume registration (article 7–1).

The submitted data vary according to types of registration undergone. In general, it contains physical and chemical properties, toxicology, exposure and hazard assessment. Substances are evaluated and based on hazard properties, toxic chemicals are divided into four classes (Article 3):

- **Class 1 chemicals** (107 substances) cover those chemical substances that are not prone to decompose in the environment or that pollute the environment or endanger human health due to bioaccumulation, bio concentration or biotransformation.

- **Class 2 chemicals** (100 substances) covers those chemical substances that cause tumors, infertility, teratogenesis, genetic mutations or other chronic diseases.
- **Class 3 chemicals** (72 substances) covers those chemical substances that endanger human health or the lives of biological organisms immediately upon exposure; and
- **Class 4 chemicals** (92 substances) those chemical substances for which there is concern of pollution of the environment or the endangerment of human health.

Chemical management is based on the class. However, for all classes companies must regularly report records concerning the handling of toxic chemical substances according to Article 8. Marking and labelling of containers and packages together with preparation of safety data sheets is compulsory (Article 17).

In the case of class 1, 2 and 3 chemicals there is obligation to submit risk prevention and response plans (Article 10). Application for a permit, registration of use or storage or approval for usage below than large-scale usage may be required under Article 13. There is also obligation to employ professional technical management personnel (Article 18). Class 4 chemicals requires submission of toxicological information and companies need to obtain an approval prior handling (Article 7). Usually the validity of permit, registration documents or approval is 5 years (Article 14). It is expected that these rules are currently affecting around five thousand stakeholders (He 2013).

There are two types of penalties under TCSCA related to registration and notification requirements (Article 35–1). First group of penalties covers the failure to notify new substances with possible fine ranging between TWD 200 thousand to 2 million TWD (between 6 and 60 thousand USD) and in the repeated infringement business may be suspended or export may be returned. Second group covers failure to register designated existing substances. Fines in this case ranges from 30 to 300 thousand NT. However, more acts are defined. For example according to Article 29 the convinced perpetrator cause human death shall be punished by life imprisonment or a minimum of 7 years imprisonment and may be fined a maximum of TWD 10 million (around 150 thousand USD).

Under TCSCA there is no provision mentioning only representative for foreign companies. However, it is expected that foreign companies may appoint local agent to ensure compliance with registration.

For new chemical substances there is suggested individual submission. However, there is mandatory joint submission regarding existing chemical substances and submitting companies can use data already submitted in order to avoid duplication of testing (Article 7–1). In the case of confidentiality and sensitive business data Taiwan Environmental Protection Administration (EPA) may allow individual submission.

As noted by ChemSafetyPro (2015b), there are significant differences between original EU REACH and TCSCA. From the early beginning (contrary to EU REACH) TCSCA address small quantity registration for substances produced under 100 kg and simplified registration for substances in tonnage between 100 kg and 1 ton. The scope of registration for existing substances is more complex under EU REACH as only designated existing chemical substances are registered under TCSCA. TCSCA requires registration of polymer itself in comparison to registration of monomer or reactant under REACH. The Only Representative provisions are missing under TCSCA and contrary to REACH there are no reduced data requirements for transported isolated intermediates (ChemSafetyPro 2015b).

7.4 Singapore

Chemical export from Singapore was in 2014 worth $38 billion USD. The most important export countries were China (14.7%), Belgium (9.43%), USA (8.82%) and Indonesia (7.88%) and Malaysia, which counts for 7.19% of Singapore export (WITS 2015). The EU export in chemicals counts for 18.7% (WITS 2015).

Regulation of hazardous substances in chemicals is mainly based on Environmental Protection and Management Act of 2008 which covers transport of hazardous substances, its import, its storage and supply. All mentioned activities are under supervision of the National Environment Agency (NEA). Regarding transportation the act is setting consignor's responsibility, information requirements, rules for documentation, responsibilities and instructions for drivers and labelling standards (see Part II).

Import of hazardous substances is regulated by only two provisions. Fist provisions sets out the duty that import license owner has to ensure that a substance in the tank or container is designed, constructed, maintained and labelled properly (Article 15). Because Singapore is a small country, there are specific places where imported hazardous chemicals may be collected. Sea, road, rail and air ways of entry are specified (Article 16). The act specifies also rules for storage (Part IV) and sets out the obligation to establish emergency action plan to deal with accidents (Article 22) and the obligation to report accidents, including loss or theft of hazardous substances (Article 23 and 24).

A very important part of hazardous chemicals regulation is the Management of Hazardous Chemicals programme which aims to prevent and control chemical hazards at workplace and safeguard health and safety of persons against chemicals hazards (WSH Council 2011: 4). There are 12 key elements of the programme which should ensure a high level of protection. Among the most important points belongs the proper selection and procurement procedure when all new chemicals should be evaluated for hazards before acquisition or purchase (WHS Council 2011: 7). For this purpose information from Safety Data Sheets are key indicators of hazards. The system also requires a register of all hazardous substances to be kept in order to keep information about amount, location, supplier etc. (WHS Council 2011: 8). Some parts deals with signs, labelling requirements, risk assessment and control measures, safe work procedures and personal protective equipment, medical issues, waste disposal or training (WHS Council 2011: 9).

Chemical regulation in Singapore is aimed at safe handling of chemicals, rather than addressing the complex management of chemicals, including assessment of all chemical substances. In this sense we can claim that chemical management policy is specific due to fact that Singapore is small island country. Most of the chemicals are for export or re-export rather than for domestic consumption which seems to be well regulated.

Chapter 8
Exotic Dimension of REACH

This chapter explores four other Asian countries, noticeably Malaysia, Indonesia, Thailand and Philippines and their relationship with the chemical industry. All countries in the region experienced updates in their chemical regulations. Therefore the question is posed to what extent is their chemical regulation similar and to what degree EU REACH was reflected during domestic reform process?

8.1 Malaysia

Chemical export from Malaysia was in 2014 worth of 9.6 billion USD. Top destinations for export were China (15.3%), Indonesia (10.49%), Singapore (9.11%), Thailand (8.3%) and India (7.13%). EU counted for 8.2% (WITS 2015).

Malaysian chemical regulation was based on so called "Class regulation" on classification, labelling and safety data sheets for the use of hazardous chemical in the workplace and industry. The regulation was implemented by the labor ministry. It is based on the UN Globally Harmonized System and required manufacturers, importers and formulators to classify (Part II), label (Part IV), package (Part III) and fill out safety data sheets (Part V) and report all necessary data to specific inventory (Part VI) under the Department of Occupational Health and Safety (DOSH). According to the regulation the inventory of hazardous chemicals shall contain the product identifier, the name of the substance, the composition of ingredients of a hazardous chemical and hazards classification and the total quantity of each hazardous chemical imported or supplied (Part VI.14.3).

Class regulation does not apply to radioactive material, waste, cosmetics, pesticides and research and development (Part I.3) and does not generate enough information regarding new chemicals or chemicals placed on the market. The information gap was present also in Malaysia. For this reason, new legislation was drafted.

O. Filipec, *REACH Beyond Borders*, DOI 10.1007/978-3-319-54154-9_8

The Environmentally Hazardous Substances Notification and Registration (EHSNR) was drafted and entered into force in 2011. According EHSNR notifications are required on all environmental hazardous substances (included in Environmentally Hazardous Substances Reference List (EHS),[1] CMR Reference List or assigned GHS classification), also included in mixtures and products produced or imported above 1 tonne per year.

Similarly to other existing chemical regulations there is a distinction between particular types of substances. There are substances placed on the EHS Reference list[2] and other substances which must be assigned a hazard category. The notification requirement does not apply to substances where information is submitted under The Pesticide Act 1974 (Act 149), The Poisson Act 1952 (Act 366) and Chemical Weapons Convention Act 2005 (Act 641). However, it is expected that there are more than one thousand potential notifiers (Wang and Ghazali 2010: 12).

Notifications shall be submitted by a manufacturer of an EHS substance or importer of an EHS substance (HSD 2012: 3). For importer there is an option to appoint "third party representative" which is entitled to submit notification. To have Malaysian legal entity is not required. Notification is submitted to Department of Environment of the Malaysian Government. Types of notification and data submitted vary according to the type of substance.

Under basic notification are notified substances on the EHS reference list and manufacturer or importer is submitting only basic information such as substance identification, year of notification, EHS occurrence, annual tonnage, concentration range or country of export. Much more information are required for substances not on the EHS reference list. These substances are subject of detailed notification which next to the basic information includes also physical and chemical properties, EHS identification, physical hazards, hazard to human health and to aquatic environment (HSD 2012: 6).

Notification of substances is subject of deadlines which ranges based on the production/import volume. Similarly to REACH, there was voluntary notification and a pre-registration period until 2010 (HSD 2012: 4). Substances with a volume of 100 tons per year and above shall have been notified first in 2011 and 2012. This deadline is also compulsory for all substances under GHS classification as having aquatic toxicity, are suspected or known as human carcinogens, mutagenic or reproductive toxicants. Substances produced in quantities between 10 and 100 tons are subject of notification for 2013 and 2014, substances produced/imported in the volume from 1 to 10 tons are scheduled for 2015 and 2016 and substances in volumes between 100 kg to 1 ton are scheduled for 2017 (HSD 2012: 10–11).

[1] EHS (Environmentally Hazardous Substance) is defined under Section 2 of the Environmental Quality Act of 1974 (Act 127) as „Any natural or artificial substances including any raw material, whether in a solid, semi-solid or liquid form, or in the form of gas or vapour, or in a mixture of at least two of these substances or any living organism intended for any environmental protection, conservation and control activity, which can cause pollution„.

[2] The EHS Reference List is based on European Annex VI to the Regulation (EC) No 1272/2008 (Awang and Ghazali 2010: 5) which highlights the normative dimension of the EU norms.

There are many exemptions under EHSNR requirements. For example substances not included in the EHS Reference List and not classified as hazardous according to GHS, mixtures, impurities and by-products, transported intermediates, substances produced for test marketing, research and development substances, substances produced in volumes less than 100 kg per year, substances in naturally-occurring materials etc. (HSD 2012: 3).

The above presented chemical regulatory system lead to unsatisfactory results. The lack of a unitary act covering chemical management created a fragmented framework and the data gap improved only slightly (Niemec 2015a). There was also fear of data duplication by both ministries and burden on companies which had to submit the same data twice. Malaysia decided to drop registration obligation under and decided that registration will remain voluntary to avoid duplication of rules by two ministries (Roberts 2012b). Malaysia is considering reform which is more consistent with the existing regulatory system. For this reason Malaysia presented in October 2015 their 11th Plan for Chemical Management for the period 2016–2020. The plan proposes (Niemiec 2015a):

- Conduct a feasibility study on a framework for a chemicals management act, and a further study on the establishment of a national chemical management board or commission;
- establish a national chemical inventory, to include information on the quantity and usage of chemicals, and to streamline existing agencies' database system infrastructure;
- conduct and report on risk assessment of priority chemicals, based on targeted volumes, risks and use-related information;
- develop risk management measures to mitigate risk of exposure, including the phase-out of hazardous substances;
- conduct training on topics including toxicology, hazard, exposure and risk assessment; and
- develop an active dialogue between chemical industries and government regulatory bodies, including incentives and economic packages to encourage industry participation.

The status of chemical regulation in Malaysia is not satisfactory and compared to other countries in the region (Japan, Korea or Taiwan) there are existing considerable deficiencies. While the purpose of EHSNR was addressing most of the existing problems including the lack of information, it seemed to be inconsistent with existing framework which lead to doubts and later rather also voluntary application. The positive is that government is not beginning from the "green field" and concerned ministries have already regulatory experience. On the other hand substantial reform is needed and there is a lot of work to do. It will be necessary to look at the competencies of the involved ministries or even consider creating new regulatory agency. Reshaping the "old" inventory will be necessary in order to create basis for effective chemical management.

The quality of information is the key for further chemical assessment. Inspiration is in other countries with similar problems and even EU REACH may be significantly important in the technical aspects of the regulation. It could be expected that Malaysia will not follow revolutionary REACH approach and adopt registration of all existing chemicals, however REACH already revealed quantum of data which might be used. As proposed in the 11th Plan for Chemical Management, it is important to develop dialogue with chemical industry and work together to create complex and transparent environment for the management of chemicals in order to create an attractive regime which enhances competiveness, research and development on one side and high level of protection on the other side.

8.2 Indonesia

Indonesian chemical export in 2014 counts for $ 5.11 billion USD. Country exported its chemicals mainly to China (19.96%), Singapore (7.11%), Thailand (6.87%), Malaysia (6.16%) and the USA (5.32%). The EU export counted for 9.2% (WITS 2015).

The chemical regulation in Indonesia is based on several laws including Environmental Protection and Management Act No. 32/2009, Government Regulation No. 18/1999, Juncto Government Regulation 85/1999 on hazardous Wastes Management and Government Regulation No. 74/2001 on Hazardous Substances Management. The main objective of the last mentioned regulation is to prevent and reduce impact and risk of hazardous substances to the environment, human health and well being. It is based on the principles of toxic substances minimization, integrated management covering production, storage and usage and sustainable development (Purwandari 2014).

For the purposes of regulation there is a definition of hazardous and toxic material which shall be substance, energy and/or other components which due to their nature or concentration and/or quantity, both directly and indirectly, can pollute and/or damage the environment, health as well as the continuation of human life and or other living creatures (Article 1). Hazardous toxic material may be classified as explosive; oxidizing; extremely, highly flammable or flammable; extremely, highly or moderately toxic; harmful, corrosive, irritant, dangerous to environment, carcinogenic, teratogenic and as a such may be permissible for use, prohibited for use or subject of limited use (Article 5).

There are several exceptions from the regulation including radioactive materials, explosives, oil and gas processed products, food and beverages (including additives) medial supplies, cosmetics, pharmaceutical ingredients, narcotics and chemical and biological weapons (Article 3).

There are both a registration mechanism (Article 6) and a notification (Article 7 and 8) mechanism included. Registration is required for producer or the importer of hazardous substances (called "B3" as abbreviation from Indonesian Bahan Berbahaya dan Beracun) which are intended to be marketed in Indonesia. At the first stage of registration the applicant is providing data which are later assessed for complexity. In the case that some data is missing, the competent authority has right to ask additional information. After Administrative verification technical verification follows. After successful technical verification there is "field verification" and later the approval document is issued, including registration number. Registration is uploaded into INSW (see Chap. 3 for further details). In 2014 there were 2171 registrations in Indonesia, 84% of them were successfully completed and 8% disapproved or not registered (Purwandari 2015: 10).

Notification in the form of prior consent from the authorities is required for exporter or importer for other than B3 listed chemicals. After necessary documents are sent for notification, they are evaluated and notification letter is sent to notifying company, requiring response. This response is evaluated and based on information submitted there is explicit consent given (Purwandari 2015). Additional requirements are set for production, transportation, distribution or storage regarding symbols and labelling (based on Ministerial Decree No. 03 year 2008 on Procedure for Providing of Symbol and Label on Hazardous and Toxic Substances). Currently this system reflects GHS, which is mandatory for single substances and since 2017 will be mandatory for mixtures.

There are only 209 substances on the list of toxic and hazardous materials, which are in the category of usable substances under Annex I of Regulation No 74 (they need only registration to be imported). There are another 10 substances under Annex II, Table I, which are limited in use based on a notification permit. Forty-five substances are subject of Annex II, Table 2, which are banned to import and use (Bahri 2011: 5).

This according to Ahman Ashov Birrry from Greenpeace does not correspond with the amount of harmful chemicals on the market (Birry 2014). Nevertheless, this is not the only problem in Indonesia as it seems that the industrial lobby campaign was successful and the Ministry of Environment turned in favour of the industry. The draft chemical legislation only applied to new chemicals and not on chemicals already on the market and allowed dumping of hazardous wastes into the sea (Birry 2014). Moreover, some companies are not reporting the import of hazardous chemicals (Purwandari 2015: 17).

The above mentioned shortcomings may be addressed by enhancing capacity of relevant authorities which may use resources to extend control mechanism and apply sanctions. Another point is activity in the international networks which may lead to improvement in legislation when knowledge will be transferred in to the proposal of amendment. The government may need to review the list of toxic and hazardous materials, as the number of chemicals included is very small. Again, EU REACH and generated data may be a useful guideline to consider the relevancy of certain chemicals to be strictly regulated. Considering REACH data may be promising, as there are only 209 toxic substances registered in Indonesia but more than 1000 substances under REACH Annex XVII.

8.3 Thailand

Chemical export in Thailand was during 2014 worth of $12.2 billion USD. Most important countries for Thai export were China (23.3%), Japan (9.31%), Indonesia (7%), Vietnam (5.8%) and India (5.4%). EU counted for 5.3% (WITS 2015).

Chemical regulation in Thailand is based on Hazardous substance Control Act (HSCA) which entered into force on 7 April 1992, touching all entities engaged in manufacturing, import, export or possession of chemicals. HSCA has been amended in 1995 and 2008. Hazardous substance is considered to be explosive, flammable substance, oxidizing agent and peroxide, toxic substance, substance causing diseases, radioactive substance, mutant causing substance, corrosive substance, irritant substance or other substance either chemicals or otherwise which may cause injury to the persons, animals, plants, property or the environment (Section 4).

The basic element of the regulatory system is the Hazardous Substance List and its annexes. This database was created to comply with the GHS classification criteria (GHS Rev. 3) and is maintained mainly by the Department of Industrial Works and other relevant authorities. The list consists of 6 annexes, including in total over 1500 entries (Niemiec 2015b). Each annex covers chemicals registered by a different ministry under different law and is sectoral in its nature.

Annex I is filled by notifications under the Department of Agriculture and includes mainly pesticides. Annex 2 is administered by the Department of Fisheries and includes chemicals used in biocides for the purposes of controlling, preventing, and destroying microorganism, parasites, plants or other animals in fisheries and aquatic animal farming. Annex 3 is administered by Department of Stock Development and covers mainly veterinary drugs, disinfectants and other products used for sanitation. Annex 4 deals with chemicals used for public health and is administered by Food and Drug Administration. Annex 5 is administered by the Department of Industrial Works. It seals with controlled chemicals substances with clear identifiers, wastes, used electrical and electric appliance, other substances such as hydrochloroflourocarbons, chemical weapons and since 2015 also all other chemicals meeting the definition of hazardous substances under Hazardous Substances Act. Annex 6 is administered by the Department of Energy and covers natural gas and liquefied petroleum gas. It means that some chemicals are registered under other annexes than Annex 5.6 and thus are subject of exception.

Information submitted during notification reflects the information of Safety Data Sheets: legal entity info, type of activity, the name of the substances and HS code, type of the substance or mixture, composition, UN number and class, GHS classification, physic-chemical properties, toxicological info, eco-toxicological information and disposal considerations etc.

There are four classes of chemicals which requires different obligations based on the degree of hazard identified. The more hazardous class, the more obligations present (Section 18):

- **Type 1 chemicals** require notification of the product information and compliance with specified rules under labelling and rules for manufacturing and storage;
- **Type 2 chemicals** require registration and notification to the authority on the action taken for production, import, export or possession and notification of quantity of production, export, import or possession;
- **Type 3 chemicals** require registration and subject must apply of a license to produce, import or posses such hazardous chemical. There is also notification of quantity production and other requirements similar to type 2 substance;
- **Type 4 chemicals** are prohibited for production, export, import or possession.

The management of hazardous substances vary according to operation and type of the substance. For this reason there is a number of supporting acts, such as:

- The Notification of MOI B.E. 2551 (Designation of a Specialized Person Responsible for Safety of Hazardous Substance Storage under Authorization of Department of Industrial Works at the Hazardous Substance Business Facility).
- The Notification of DIW B.E. 2551 (Criteria and Method for Notifying and Registration of Specialized Person Responsible for Safety Storage of Hazardous Substance under Authorization of Department of Industrial Works and Reporting of Safety Storage of Hazardous Substance).
- The Notification of DIW B.E. 2553 (The Examination for a Registration of Specialized Person Responsible for Safety Storage of Hazardous Substance).
- The Notification of MOI B.E. 2555: Hazard Classification and Communication System of Hazardous Substances)

Next to the Hazardous Substances Act supporting legislation, there is also Ministerial legislation setting up license fee and duties of persons engaging in the activities with individual types of chemicals. There are also Notifications of the Ministry of Industry which deals with separate categories of chemicals (wastes), its transportation, registration etc. Similarly there are notifications of Department of Industrial Works and Regulations of the Department of Industrial Works, manifest Forms and relevant manuals. In total, there are around 35 other documents (see Jetro Thailand 2015).

The last modification comes by the B.E. 2558 on the notification of the production and import of hazardous substances on the 19 February 2015. There was new "sub-annex" 5.6 created on the Thailand Hazardous Substances List. This annex includes substances classified as type 1 hazardous substances (Kongprachaya 2015). According to this legislation manufacturers and importers of hazardous substances which exceeds 1 ton per year are required to submit notification to Department of Industrial Works (DIW). There is a period of 60 days from the date of manufacture or import (Kongprachaya 2015). DIW announced creation of the inventory of existing chemical substances. All substances notified before 31 December 2016 will not be listed in the inventory and treated as new chemical substances (CIRS 2016). This is step enhancing registration and shifting it closer to REACH logic.

Thai regulatory system for chemicals is constantly changing. In 2016 the preliminary inventory of existing chemicals was presented, combining hazardous substances list and latest hazardous chemical notification list. Companies had opportunity to notify hazardous substances until the end of 2016. After that date, chemicals not listed in the inventory are considered new and notification will be required. Possibly, the system of prioritization for further management will be proposed. Enlisted substances and substances not matching SVHC criteria will be suitable for simplified risk assessment report. In other cases full risk assessment will be required (See ChemSafetyPRO 2016; Kongprachaya 2016). Latest news suggests, that chemicals determined to be SVHC will require chemical risk assessment reports depending on tonnage. Similarly to REACH, DIW proposes grace periods of 3, 5 and 8 years based on production/import volume (up to 10 tons, up to 100 tons and up to 1000 tons or more respectively).

It is too early to evaluate all proposed changes as the system is not fully consolidated. However, it seems that considerable parts are changed in the direction towards REACH and there is also possibility to adopt Canadian model for prioritizing assessment of enlisted substances.

8.4 Philippines

Chemical export in the Philippines counted in 2014 for $1.5 billion USD. Most important destinations were Japan (25.4%), China (21.1%), USA (6.21%) and Netherlands (5.18%). The EU counted for 10.4% (WITS 2015).

Chemical regulation in Philippines is based on the Toxic Substances and Hazardous and Nuclear Waste Control Act of 1990 (TSHNWC). The main aim of this act is to regulate, restrict or prohibit, the importation, manufacture, processing, sale, distribution, use and disposal of chemical substances and mixtures that present unreasonable risk and/or injury to health or the environment and to prohibit the entry, even in transit, of hazardous and nuclear wastes (Section 2). For this purposes an inventory (PICCS) has been created to list chemicals imported, manufactured or used to support evidence of relevant data regarding chemicals (Section 4). Under section 5 hazardous substances are defined as substances presenting short term acute hazards (acute toxicity by ingestion, inhalation or skin absorption, corrosives or other skin or eye contact hazard or the risk of fire or explosion or/and long term environmental hazards including chronic toxicity, carcinogenetic, resistance to detoxification, biodegradation, potential to pollute underground or surface waters or aesthetically objectionable properties such as offensive odors (Section 5).

This act has been strengthened by the Pre-Manufacturing and Pre-Importation Notification (PMPIN) for New Substances. However, even under section 8 of TSHNWC there was an obligation to submit necessary information regarding pre-manufacturing or pre-import if they are not exempted (listed already on PICCS, research and development purposes, not presenting unreasonable risk to the health and environment; or chemicals that exists temporarily and with no human or environmental exposure).

There are several exemptions from PMPIN, including chemical substances already included in PICCS, small scale premises, small quantity chemicals (produced under 1 ton per year), certain polymers and other specified substances, non-isolated intermediates, articles and new chemicals produced exclusively for export. However, these exemptions may be subject of specific rules and procedures (EMB 2015).

Small Quantity Substances Certification (SQI) is used for importation of chemicals matching the volume less than 1000 kg per year. Polymers of Low Concern (PLC) are subject of Polymer Exemption Certificate if all of the monomers are listed on PICCS or 2 or more of the top monomers are included in the definition of another polymer already in PICCS. Letter for request is required. Even substances listed on PICCS may be subject of voluntary certification. This may be required by the Bureau of Customs or other authorities. Letter of request is also required and submission of basic information is necessary.

Notifications and data requirements vary due to origin of the new chemical. Abbreviated form of PMPIN is used for countries with similar review procedures and the substance is already listed in the inventory of those countries. This applies to the EU, USA, Australia, Canada, South Korea or Japan. For other new substances is necessary to use Detailed form of PMPIN with some special requirements and reports including environment effects, specific uses, annual quantity produced or imported, test reports about physic-chemical properties, reports on eco-toxicity etc. (EMB 2015).

After notification is granted, the company must submit Notice of Commencement to Import or Manufacture Form. After this procedure is successfully done, the new substance is added to PICCS. If requested, it may be added to PICCS confidential part of the inventory).

Section 9 of TSHNWC requires testing in all cases where:

- There is a reason to believe that the chemical substances or mixture may present an unreasonable risk to health or environmental exposure thereto;
- There are insufficient data and experience for determining or predicting the health and environmental effects of the chemical substance or mixture; and
- The testing of the chemical substance or mixture is necessary to develop such data.

Sanctions for any violation of law are considerably high. For example person who is knowingly using chemical substance or mixture which is imported, manufactured or processed in violation of the act or implementing rules may be imprisoned from six up to 20 years. The same punishment may apply for failure to submit reports, notices or on the information to records or the failure or refusal to comply with the pre-manufacture and pre-importation requirements (see section 13 and 14). Administrative fines may apply as well (Section 15).

System used in Philippines seems to be very special, based on notifications which ensures basic data gathering. System also allows certain degree of simplification which is partly based and relies on the regulatory system of developed countries.

Chapter 9
REACH and "the Rest of the World"

In previous chapters we have focused on 10 states outside the "European neighborhood" which responded to EU chemical regulation in different ways. However, there are more states affected by REACH which had to act. Some of them are more integrated into international networks and were proactive on the international level from the early beginning (Switzerland, Canada, Australia) while others have similar attitudes like the BRIC countries. In this chapter chemical regulation in some remaining countries will be explored and at which level are they affected by REACH, regardless of their geographic position in order to improve the complete picture about EU chemical regulation on the international level. This chapter deals with the chemical regulation within Canada, Australia, New Zealand and Switzerland including other countries which decided to improve its chemical management.

9.1 Canada

Chemical export from Canada in 2014 was worth of $ 28,5 billion USD. The USA are for Canada the most important export destinations because 69% of Canadian chemicals goes there. Among other top destinations belongs China (4.86%), Japan (2.61%) and Italy (2.48%). The EU counts for 6.2% (WITS 2015) which makes EU second most important subject after USA. However, the disparity between the USA and EU is obvious.

Canada has held for a long time a very good reputation in the field of chemical management and the Canadian Environmental Protection Act of 1999 (CEPA) has been the normative model for other states seeking for high level standard of chemical regulation. Due to its precautionary approach (Part 1), phasing-out approach to hazardous chemicals and data gathering provisions (Part 3) may be well compared to EU REACH. A review every 5 years provides the opportunity to improve the performance of the agency and flexibly react to actual challenges. However, CEPA is not aimed only at regulation of chemicals. It covers also vehicles, living organism,

O. Filipec, *REACH Beyond Borders*, DOI 10.1007/978-3-319-54154-9_9

wastes or maritime environment. A great part, however deals with chemicals and especially toxic chemicals.

For the purposes of the regulation a substance is toxic if it is entering or may enter[1] the environment in a quantity or concentration or under conditions that (Section 64):

- (a) Have or may have an immediate or long-term harmful effect on the environment or its biological diversity;
- (b) constitute or may constitute a danger to the environment on which life depends; or
- (c) constitute or may constitute a danger in Canada to human life or health.

Next to the CEPA there are other pieces of legislation which covers specific chemicals. These includes the Hazardous Products Act establishing standards for classification, hazard communication and regulation of chemicals, Pest Control Product Act, Food and Drugs Act, Transport of Dangerous Goods Act, Fisheries Act and Canada Labour Code.

A very important tool is the Chemicals Management Plan (CMP) which is the initiative of the Government of Canada aimed at reducing risk posed by chemicals and setting up the priorities and deadlines for actions to be taken on chemicals of concern. The plan is complex and includes also priorities in integrating management activities among federal departments, support of research, strategies to increase industry stewardship, international cooperation and communication with downstream users.

The plan is divided into three phases and each is aimed at different set of activities. In the first phase of CMP (2006–2012) the plan focused on activities entitled "Industry Challenge", "Rapid Screening" and "Petroleum Sector". During the Industry Challenge the Industry had obligation to supply additional data for cca 200 substances identified in the categorization as the highest priority and when proven necessary to adopt risk management. Rapid Screening covered potential persistent, bioaccumulative or pontetially toxic substances of which some required further assessment. Similarly, last gathering initiative focused on pertoluem industry which when applicable was required to propose risk management options related to some substances. Second phase (2012–2016) continued with further "Rapid Screening" based on exposure-related information, "Substance Grouping Initiative (SGI)" and measures focusing on polymers. During SGI additional 500 substances were put into nine similar groups and proceed similar assessment to Industry Challenge. As for polymers, there is proposal for information gathering in several phases. For the third period scheduled between 2016 and 2020 there is a need to assess approximately 1700 substances for the level of risk (for more details about CMP see Abelkop and Graham 2015: 42).

[1] Substance may become CEPA-toxic based on a showing of potential harm under the evidence of significant hazard or exposure, not necessarily both. This attitude is similar to REACH precautionary principle (see Denison 2009b).

CEPA provisions are setting out that new substances manufactured or imported into Canada since 1994 in annual tonnage higher than 100 kg must undergo human health and environmental assessment (Section 81). Based on the results of assessment, regulatory measures may take place. It is estimated that the Government of Canada assesses approximately 600 substances new to the Canadian marketplace each year.

CEPA is based on several elements. First element is the inventory. Canada has created an inventory called Domestic Substance List (DSL) which makes distinction between new substances and existing substances (Section 66). As noted by Anne Wordsworth (2007) the list is composed of chemicals marketed between 1 January 1984 and 31 December 1986 (Wordsworth 2007: 7). There are approximately 23,000 existing substances marketed in Canada which are subject of examination to determine potential hazard to human health or environment. Next to the DSL there is also the Non-Domestic Substances List where the remaining substances are placed (Section 66). The Distinction is crucial, because substances not present on the DSL are considered new to Canada and thus subject of higher information requirements and New Substances Notification (only certain chemicals such as articles, impurities, by-products or naturally occurring substances are subject of exception).

Second element is information gathering. Under section 70 each person has an obligation to inform the Minister without delay about the fact that a substance produced, imported or possessed is toxic or may became toxic. Notification requirements may be set after publication in the Canada Gazette requiring submitting necessary information (or even samples) regarding hazard properties of concerned chemical and may also require compliance with the notice. Minister may also demand toxicological and other tests, plans or even specific actions to be taken by the subject engaged in chemical activity (Section 71).

The third element of chemical management in Canada is the "categorization" approach which is trying set up priority for assessment. Priority is given to inherently toxic substances which are harmful by its very nature to human or to the environment, persistent, bio accumulative and substances to which people might have the greatest potential for exposure (section 73). Categorization was completed in 2006 and identified approximately 4300 substances from LDS requiring further attention. Screening of substances in order to determine whether a substance is toxic followed in 2007, resulting in identification of 1200 substances of low ecological concern and 750 substances identified as potentially not of concern (see Section 74). There were 200 identified chemical substances representing highest priorities for risk assessment and appropriate controls, termed as "Challenge" chemicals.

The fourth important element is so called "Priority Substances List" (PSL) where priority substances for assessment are placed. There is a 5 year period for assessment. Placement on the list is made upon consultation with relevant scientific and political authorities and any person may demand inclusion of chemical on the list (Section 76). Inclusion of chemical on the PSL is political decision, however ministers have obligation to consider a screening assessment and scientific consideration (Section 76). Ministers may decide to take no action, to place a substance on the

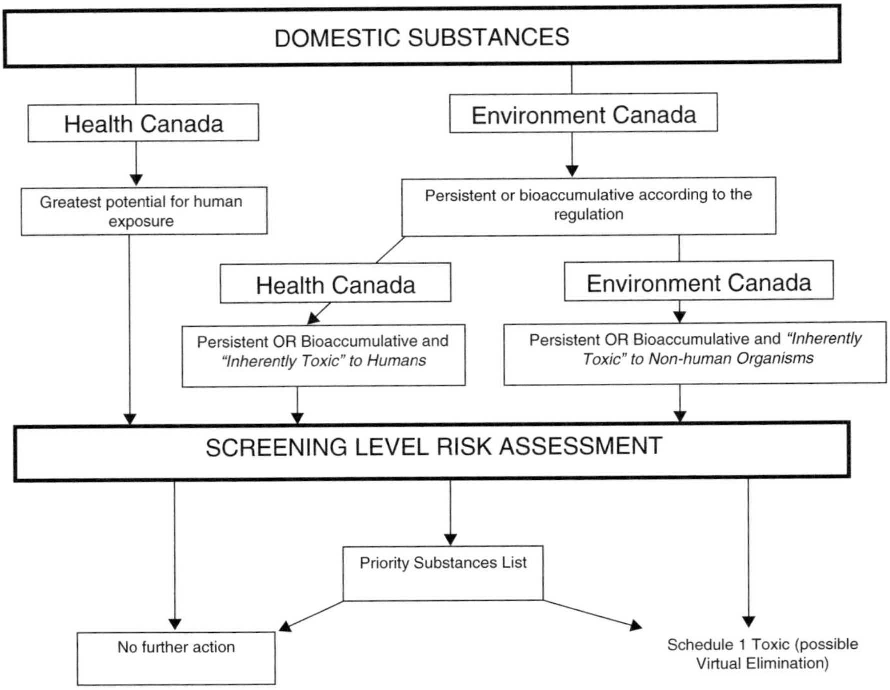

Picture 9.1 Domestic substances assessment
Source: Environment Canada (2015)

Priority Substance List (again) or in the case that the substance is toxic, recommend its placement on the Toxic Substances List[2] Schedule 1 or propose virtual elimination. The last measure is mandatory to apply when substance has proven to have inherently toxic, bio accumulative or persistent properties (Section 77). However, the scope of substances for virtual elimination[3] is not limited and other criteria may apply. Despite open requirements for this category there were for a long time only two chemicals placed: Hexachlorobutadiene (HCBD) and Perfluorooctane Sulfonate and its salts. For example, if the substance pollution results from human activity or the substance is not naturally occurring in the environment. The final decision is published as well as report from the assessment. A company affected by the decision may notify the objection that the used chemical has been placed on a Substance Priority List (Section 77). Picture 9.1 shows the procedure of domestic substances chemical assessment.

[2] To the Toxic Substances List are placed chemicals which have or may have an immediate or long-term harmful effect on the environment or its biological diversity, constitute or may constitute a danger to the environment on which life depends or constitute or may constitute a danger in Canada to human life or health.

[3] This is defined as: „*the ultimate reduction of the quantity or concentration of the substance in the release below the level of quantification specified by the Ministers in the List referred to in subsection*" (Section 65).

The process with challenge chemicals started in 2007, when the EPA required additional information from stakeholders in order to close information gap. Based on the process, most dangerous substances were identified and determined for further additional risk management. Information on remaining substances were gradually collected and DSL properly updated. However, it is important to note that other 9000 substances are assessed under the Food and Drug Act, and around 400 substances under the Pest Control Products Act. If we compare this dedication with other countries, Canada has next to the EU the second largest policy aimed at evaluation of chemicals.

Among other important aspects belongs notifications regarding every "significant new activity" in respect of substance has to be notified (it covers substances on DSL or new substances). This may include release of the substance into environment, potential exposure, significant difference in quantity or concentration production or alternative use of the chemical (section 80). Subject of which is manufacturing or importing substances not specified on the DSL and is not covered under exceptions, and then the activity is illegal. The person must notify the production or import of such substance. There is a specific waiver procedure regarding significant new uses of chemicals (see section 81) and criteria for notification are published. Failure in notification or at any case when the Minister has "reasonable grounds" to believe that person did not fulfilled its information requirements may lead to prohibition of the activity (Section 82).

Restrictions apply to chemicals on the Toxic Substances List. This may consists of various regulatory measures including regulations, specific obligations, codes of practices, special rules for storage or handling etc. Export of chemicals from Canada may be also subject of restriction or permit. There are three Parts of the Export Control List. Part 1 includes prohibited substances, Part 2 includes substances which are subject of notification or consent and some substances are restricted (Section 100).

Another important aspect of CEPA is research, monitoring and surveillance in order to identify substances, which has effects on human health and the environment (Section 44). For this purpose CMP established Environmental Monitoring and Surveillance Program. Special focus is laid at pollution prevention. For this reason EPA may collect information related to pollution and require pollution prevention plans (Part 4).

Canadian chemical regulation is also adopting slightly different attitude than REACH regarding the scope of chemicals registered and assessed as only selected chemicals are subject to deeper evaluation. However, CEPA created complex and integrated system which allows high level of flexibility. This is due to clearly distinguishable material and procedural parts. The variety of lists fully matches the procedural provision which reflects the consensual principles of the Canadian political system. For this reason it was possible to incorporate the role of regional authorities within the system, grant them certain rights during regulation and reflect their voice during consultation. CEPA system seems to be generally transparent as most important decision are subject of publication including assessment reports. Any person may report and demand restriction of some substance trough official legal way.

9.2 Australia

Australian chemical export in 2014 was worth $10.9 billion USD with the majority of unspecified origin according to WITS data. However, New Zealand counted for 7.34%, Korea for 7.31% and USA for 4.57%. It could be expected, that EU counted for less than 6% (WITS 2015). DG TRADE made evidence of Australian import of chemicals worth 482 million EUR, which is roughly 4.3% share[4] (DG TRADE 2015b: 4).

Chemical management in Australia is based on Industrial Chemicals Act of 1989 which is aimed at notification and assessment of chemicals. The act has been several times amended. For the last time by the Act 59 of 2015 from 1 July 2015. The act established the National Industrial Chemicals Notification and Assessment Scheme (NICNAS) under section 2. The main aim of NICNAS is to assess all industrial chemicals which are new and evaluate their effects on human health and the environment. NICNAS does not cover articles and radioactive materials, medicine products, pesticides, veterinary chemicals, food and food additives. In the case of mixtures only individual substances are covered by NICNAS (section 2).

The first aspect of Australian chemical management is the distinction between existing substances (listed in the Australian Inventory of Chemical Substances – AICS and new chemicals substances. Similarly to other regulatory systems, new chemical substances in the case of import or manufacture must be notified and the importer must submit New Chemical Notification prior the activity. However, only the Australian subjects (producers, importers) are allowed to submit new substance notification.

Companies are exempted from notification if they manufactured chemicals in Australia for research, development and analysis purposes; imported new chemicals to Australia solely for research, development and analysis with quantity less than 100 kg per year; chemicals which are at ports subject to customs control for maximum of 30 days; chemicals that are posing no unreasonable risk and are in quantity less than 100 kg per year; chemicals that are non-hazardous and introduced in cosmetics with concentration less than 1%. In all the cases (except that chemical is produced in Australia) an annual report is required (Section 4).

New chemical substances which are not exempted are subject of assessment. However similarly to other regulatory regimes existing chemicals were designed for assessment under the Priority Existing Chemicals (PECs). Companies may use for non-hazardous substances or priority substances self-assessment frame (28 days) or may use non-self assessment frame which may take 90 days (see Division 5).

---

[4]This is only approximate number as the methodology used may differ (e.g. DG TRADE may consider under chapter chemicals different items of goods than the World Bank).

There are several obligations for the companies regarding their activity and character of the substance. All subjects producing or importing industrial chemical for commercial purposes shall be registered and obtain business registration. It is necessary requirement to enter Australian market and thus no exemptions regarding volume produced apply. Registration is paid and fees or charges vary according to the value of chemicals manufactured or imported. Other obligations are based on the company's activity.

There are several types of permits of certificates producer or importer may apply to, based on the character of substance, the purpose and production volume. Permits and certificates also vary due to data requirements, administration time needed, publication in the AICS or in the Chemical Gazette. Permits have shorter statutory timeline, reduced information requirements and are usually limited in validity. There are four types of permits and three types of certificates (see part 3):

- **Commercial Evaluation Permit (CEP)** may be required for chemicals intended for test market purpose. Its validity is limited to maximum 4 tonnes and 2 years (Division 1A);
- **Low Volume Permit (LVP)** may be required for new chemicals produced in the volume less than 100 kg per year or maximum of 1000 kg per year if the chemical meet defined low hazard criteria (Division 1B);
- **Controlled Use Permit (CUP)** may be used for low-risk new chemicals due to high standard of use, handling and disposal control. CUP is limited only to certain activities such as export (Division 1C);
- **Polymer of Low Concern Certificate (PLC)** may be used for new polymers meeting prescribed criteria. There is no volume limit (AG 2015).
- **Limited Notification Certificate (LTD)** is easy to obtain as there are reduced data requirements. This certificate may be used for small-volume chemicals, biopolymers and low MW synthetic polymers, site limited chemicals, biopolymers and low MW synthetic polymers or synthetic polymers that does not meet the PLC criteria (AG 2015);
- **Standard Notification Certificate (STD)** is required for any new chemical produced in a quantity more than 1 ton per year and which does not meet criteria of other category of permits or certificates (AG 2015); and
- **Early Introduction Permit (EIP)** allows notifiers submitting PLC, LTD or STD to introduce new substance before assessment is completed. The EIP chemical must meet prescribed criteria and the validity of EIP is limited until a proper certificate is issued (Section 30A).

The chemical regulatory system in Australia seems to be simple, making clear distinction between new and existing domestic substances. Compared to REACH Australia decides to undergo only limited assessment of industrial chemicals. However, a very characteristic aspect of Australian regulation if its focus on the executive part of the system which is based on permits and certificates.

9.3 New Zealand

New Zealand's chemical export in 2014 is worth $ 1.96 billion USD. The most important destinations for export were USA (22.67%) and the Australia (16.5%), followed by Japan (11.49%), China (8%) and the Netherlands (5.67%). The EU counted for 14.6% which makes EU third most important destination (WITS 2015).

Hazardous substances in New Zealand are regulated by the Hazardous Substances and New Organisms Act (HSNO Act) which entered into force on 29 July 1998 for new organisms and on 2 July 2001 for hazardous substances. There is the Environmental Risk Management Authority which is responsible for the implementation of the act. The regulatory system in the New Zealand is very specific compared to other systems in the region because is rather based on chemical components than on chemicals itself.

For the purposes of HSNO Act hazardous substance is defined as any substance that can damage the environment or adversely affect human health and safety other than radioactive, ozone-depleting or infectious substances (EPA 2015a). Several substances are excluded from the HSNO Act: hazardous substances only used in an exempt laboratory, medicine, infectious substances, ozone-depleting substances, food and manufactured articles. Hazardous substances cover both single component substances and formulated products (EPA 2015b).

Any person who wants to introduce new hazardous substance on the New Zealand market must apply to the Environmental Protection Agency for Individual substance approval or Group standard approval. The type of the approval depends on the composition of the product and its hazard properties. Individual substance approval is used for new hazardous chemical component which is not listed on the New Zealand Inventory of Chemicals and Pesticides (NZloC). This inventory is quite unique because it lists hazardous chemicals allowed as components in products covered by the group standard approval. If there is a new (not listed in NZIoC) single component hazardous chemical added into the formulated product or is used on its own then new chemical notification is needed. Single component chemical substances that are not hazardous do not require approval (EPA 2015a).

The second type of Group standard approval is used for formulated products and hazardous single component chemicals. Usually similar types of chemicals are approved and for the approved group are set standards to be matched (labelling, packaging, SDS etc.). The risk assessment, assigning group standards to the products and compliance with the control requirements specified by the group standards is the duty of manufacturer or importer (EPA 2015c). If the company has a problem to choose the correct standard, then it may request Environmental Protection Agency by the "Status of Substance" request (EPA 2015a).

9.4 Switzerland

The case of Switzerland is very interesting due its geographic position "in the middle of the EU" without being EU member. For this reason Switzerland is the non-EU country (except countries of the European Economic Area) which has been from the early beginning strongly affected by REACH. REACH has strong implications for the Swiss government and Swiss companies. Around 60% of Swiss chemical export goes to the EU and the EU import counts for 85%, which means that about 9000 chemicals will fall under REACH registration (Chemical Watch 2008). The most important Swiss export countries are Germany (15.38%), USA (15.35%), Italy (7.12%), France (5.87%) and UK counts for 5.42% of Swiss export. In 2014 Swiss export counted for 89.6 billion USD (WITS 2015).

For this reason the Swiss government reflected REACH in their Chemicals Act and PARCHEM Ordinances in 2005, 2 years before REACH entered into force. When REACH entered into force another pressure on Switzerland occurred and led country to consider further harmonisation of REACH. Three options were assessed: PARCHEM minus (partial adaptation), REACH Minus (partial adaptation) and REACH Switzerland, fully harmonized with the EU REACH (FOEN 2007: 6).

Switzerland revised its own legislation already on 1st February 2009. Chemicals Ordinance on Protection against Dangerous Substances and Preparations (ChemO) adopted REACH-like registration requirements on substances. ChemO deals with new substances and hazardous substances and preparations.

New chemical substances are defined for the purposes of regulation as substances not listed on the European Inventory of Existing Commercial Chemicals Substances (EINCECS) which records all substances present on the European Community market between 1st January 1971 and 18th September 1981. New substances may be subject of notification or declaration unless they are exempted. It has to be noted, that definitions are based on the EU legislative (see Art 2) which makes the act at its essence Europeanized.

Notification is used for new chemicals substances produced in a quantity higher than 1 tonne per year for the purpose of pre-market approval. Manufacturer, importer or only representative is required to notify (Art. 16) the substance to the authority and submit necessary information including physic-chemical data, toxicology, eco-toxicological data and chemical safety report (Art. 18a). The scope of data required varies according to tonnage covered).[5] Notification under Swiss regulation is identical to EU REACH registration. Substances are exempted if chemical substance is not covered by the ChemO but different acts (medicine, waste, transit regime, food additives) or is produced in a quantity under 1 tonne per year (Art. 1). Substances used as raw materials, active ingredients or additives in food, therapeutic products, animal feed or are part of substances that are used exclusively as active ingredients in crop protection products and biocidal products are also subject of exemption

[5] Swiss authority uses IUCLID 5 format for dossier submission which is another important similarity with EU REACH.

(Art. 17). Reference is made also to REACH Annex V which includes by products or naturally occurring substances.

Declarations (according to Art. 25) are used for new chemical substances produced in a quantity higher than 1 tonne per year for product and process oriented research and development (co called PPORD). In this case declaration are very similar to PPORD notification under EU REACH and the manufacturer, importer or the only representative is required to submit information regarding substance identity, uses and expected quantity at least 30 days before the activity takes place. The scope of limitations required is smaller than under the notification. However, declarations may be used only for the purpose of the product and process oriented research and development in quantities no greater than those necessary for the stated purpose and only for a period not exceeding 5 years.

Substances with hazardous properties are subject to registration (Art. 64) unless exempted. Exemptions apply to substances which are subject to other laws than ChemO as mentioned earlier or are intermediates, fall under substances and preparations solely used for research and development, are obtained in Switzerland, fall under preparations produced in the quantity less than 100 kg per year and intended for professional users or fall under gas mixtures which consists only of registered gases (Art. 69). Chemical Registration is used for substances meeting CLP classification criteria, PBT or vPvB substances, substances present on CHemO annex 7 which correspond with SVHC candidate list under REACH. Registration apply also to preparations containing at least one SVHC, PVT or vPvB substance in a concentration higher than 0.1% by weight. Different volume rules apply for gasses containing hazardous ingredient, which are contained in preparations (Article 52).

It is kind of a post-market reporting for substances and preparations with hazardous PBT or vPvB properties or are present on Swiss SVHC candidate list. Manufacturer, importer or the only representative has to register the product and submit the identity of the substance information, classification and labelling, report uses and quantity and submit chemical safety report. The registration is quite simple and provides the Swiss authority necessary information for the classification and labelling. The process is very similar to EU CLP regulation.

Additional regulation may apply to SVHC as they are also subject of Chemical Risk Reduce Ordinance (ORRChem). ORCHem prohibits and restricts marketing and use of selected substances with hazardous properties, sets out necessary qualifications for manipulation with hazardous substances etc. It is very similar to REACH annexes regarding restriction (annex XVII) and authorization (annex XIV).

Despite Switzerland not deciding for full adoption of EU REACH, ChemO fully reflects considerable parts of REACH. This is most visible in technical elements of ChemO annexes and registration provisions. It is not surprising, as Switzerland despite being outside the EU is well integrated into economic cooperation with the EU.

9.5 Other Countries

There are several other interesting countries with noticeable regulatory policies regarding hazardous substances. Countries on the Arabian Peninsula (Saudi Arabia, United Arab Emirates, Qatar and Oman) have prior licensing requirements for activities which may have negative effects on the human health or environment. This is also the case of Egypt. The Arab Republic of Egypt exported in 2014 chemicals worth $3.1 billion USD, mainly to France (16%), Turkey (12.9%), Italy (6.4%) and Spain (6.1%). Most of the export goes to the EU (WITS 2015). Trade structure makes Egypt hot candidate for possible Europeanization.

An interesting actor in the Middle East is Israel who enacted its "candle to grave" chemical control in 1993. The Israeli Ministry of Environmental Protection is responsible for the comprehensive management of hazardous chemicals in all activities including import, export, manufacturing, packaging, usage etc. Despite complex regulation, Israel does not plan to adopt GHS (Chemical Watch 2009). Israel's total export of chemicals in 2014 was worth $16 billion USD. Due to its international isolation in the region most of the export went to the USA (23%), the UK (14.5%) and Turkey (12.5%). The EU counts for almost 43% (WITS 2015) which makes Israel one of the most EU dependent countries regarding chemical export. This also makes Israel possible adept of Europeanization in the field of chemical management in the future.

On the other side of the African continent is the most important market located in South Africa. Despite most of the chemical export worth in 2014 almost $5.7 billion USD went to the USA (12%), among top 5 export countries is also Belgium (6%). However, South African export is mainly Sub-Saharan Africa oriented (WITS 2015). Unfortunately, Sub-Saharan Africa due its political problems as well as other parts of Africa is ill suited for Europeanization of chemical management policies due to financial, administrative or technical reasons.

Chemical legislation changed also in Mexico where Mexico's National Institute of Ecology (INECC) published its National Inventory of Chemical Substances (INSQ) which provides basic information on chemicals present on the market since 2009 (Chemical Watch 2013). The creation of a specialized database on hazardous substances is being considered as the country would like to proceed in the implementation of the UN Strategic Approach to International Chemicals Management (Chemical Watch 2013). However, Mexico is also not primarily oriented on Europe as almost 40% of its exports goes to the USA and American continents count for almost 80% of Mexico's chemical export, which in 2014 was worth of $11 billion USD (WITS 2015).

Part III
Towards Global Regulation?

Chapter 10
The Implications of REACH

There are many implications of REACH, both practical and theoretical. Adoption of REACH in the EU influenced international trade with chemicals and introduced new normative rules and principles on the international level. Thus REACH has normative implications for other countries, especially countries trading with the EU and practical rules for companies within non-EU member states. This normative dimension may be considered a supporting factor of adoption REACH-like regulation in the country trading with the EU. Is there a direct link between the chemical export and reflection of REACH? How can we measure the impact of REACH? In the next section practical implications of REACH are presented and evaluated. In the Sect. 10.2 theoretical implication are considered.

10.1 The Impact of REACH

The regulation of hazardous chemicals overview presented in Part II indicates that expectations about REACH influence are rather exaggerated. No country in the World has the same regulation as REACH and even Switzerland which is strongly dependent on the EU economically, there is strong interdependence as well on the political level, has chosen in some aspects sligthly different approach to hazardous chemicals management.

The hypothesis that more export to the EU (and the necessity of compliance) the more REACH-like regulation seems to be only of little relevance despite correct logic. International trade with chemicals is an important aspect of interdependence because countries that trade with each other have a tendency to enter into agreements and create common rules. REACH may affect this dimension significantly as there is an obligation for importers under REACH to comply with this EU regulation in data gathering provisions relevant to non-EU manufacturers. In the case of data submission failure their products would be not allowed to enter the EU market under "no data, no market" principle. This fact has significant implication for

O. Filipec, *REACH Beyond Borders*, DOI 10.1007/978-3-319-54154-9_10

Table 10.1 Import share and export value

State	Export to the EU (% share)	Chemical export value (bil. USD)
USA	30.5	164
Brazil	20.7	11.4
Russia	29.5	21.4
India	22.4	34.9
China	15.6	109
Japan	10.6	54
South Korea	4.3	38.4
Taiwan	6	1.2
Malaysia	8.2	9.6
Indonesia	9.2	5.1
Singapore	18.7	38
Thailand	5.3	12.2
Philippines	10.4	1.5
Canada	6.2	28.5
Australia	4.3	10.9
New Zealand	14.6	2
Switzerland	60	89.6

Source: Author, based on WITS 2015 data

assessing normative influence of REACH and EU chemical legislation. Because companies have to follow with EU data requirements, it is easier for the non-EU governments to introduce EU-like provisions within domestic regulation because the costs of introduction are reduced by EU standards. Thus, countries with a higher EU import share (Table 10.1) in chemicals should have a more Europeanized chemical regulation, at least in data gathering provisions and related hazardous chemicals regulation.

However, countries with a high share of EU export only rarely reflected REACH within certain parts of their own legislation despite the initial consideration. Most of the countries (USA, Russia or China) have rather chosen their national way to improve existing regulation. Other countries, especially India and Brazil, seems to have little experience with complex regulation of industrial chemicals and suffer many problems including lack of experience, capacities or weak infrastructure and administration. Inspiration by REACH seems to be an expensive adventure for their developing market.

On the other hand, it seems that the level of export reflects geographic conditions. This is the case of the USA and Russia. Interesting is also dependence of BRIC countries on EU export which is similar, ranging from 29.5% (Russia) to 15.7% (China). There are two states which are surprising. Due to similar geographic position it could be expected that Canadian dependence in chemicals export will be similar to USA. However, the majority of Canadian export goes to the USA. On the other hand New Zealand is on the other side of the globe than the EU and New Zealand exports 14.6% of its chemicals to the European Economic Area. This is

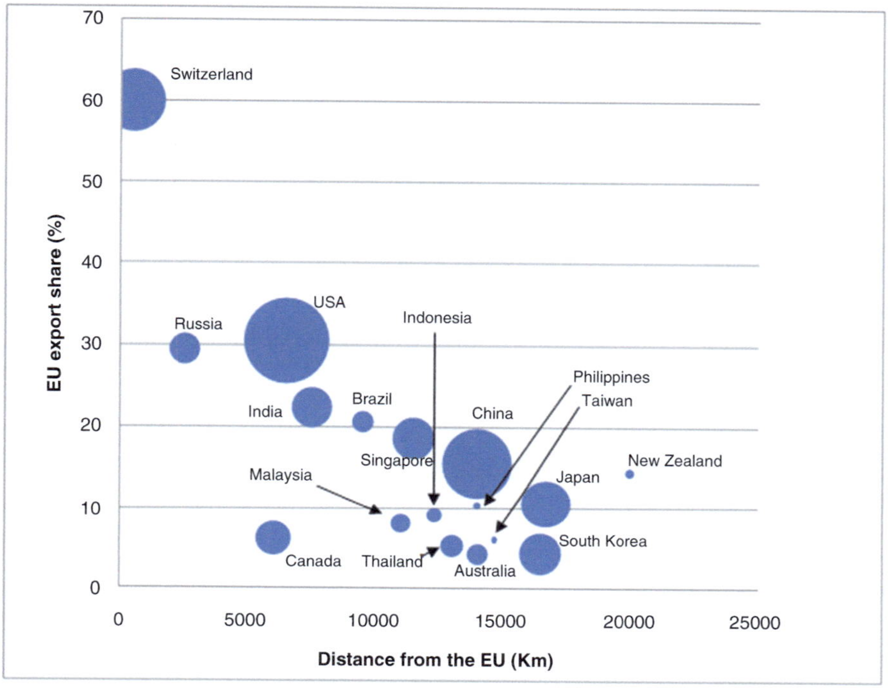

Chart 10.1 Geography and EU export share
Source: Author, based on WITS 2015, own calculation
Note: for the distance calculation naval trajectory has been used except Switzerland and Russia. Naval distance has been measured between the relevant national cities and two EU ports: Rotterdam (USA, Canada and Brasil) and Port of Gioia Tauro (for Asia, Australia and New Zealand)

more than Malaysia or Indonesia which are a thousand miles closer. However, geography matters also here, because there are several strong chemical markets including China, Japan or India. In 2014 the USA imported chemicals worth $ 183 billion USD. In comparison, China imported chemicals of $ 124 billion USD, Japan for $ 55 billion USD and India for $ 38 billion USD (WITS 2015). The three mentioned states alone have a higher import than the USA. It seems logical that countries in the region focus on access to these three countries. Chart 10.1 presents the relationship between geographic position and EU share on chemical import. The size of each point represents relative size of the total chemical export in 2014.

Chart 10.1 provides visual indication of the EU influence in the terms of EU import share and the relevant distance. There are two important elements visible from the chart. First, geography matters. There is visibly demonstrated clear tendency that with the increasing distance the share of the EU on the country export decreases. Expressed by Pearson correlation coefficient R is −0.73 which is negative moderate correlation (distance in km is independent variable, EU export share as dependent variable). This correlation would be possibly stronger if we take into the count development level of the countries considered, structure of chemical industry or regulatory history.

Second, there is a group of Asian states which are close to each other with the dominant point representing China. This may help us to understand similarity of Asian regulatory regimes. The presence of strong actors in the region may be one factor undermining the importance of REACH. However, we cannot say that REACH had no impact at all. It is not only coincidence that almost all countries revised its chemical regulation in the last few years which resulted in the implementation of similar aspects into the national regulations.

However, most of the regulations require data submission which is usually risk based rather than based on tonnage (Simpson 2012). Also testing and laboratory standards may differ significantly. In this sense OECD Mutual Acceptance of Data programme which require compliance with OECD Test Guidelines and CLP is a very promising platform harmonizing standards in countries like Canada, Japan, Korea or Turkey (Simpson 2012). Many countries incorporated "only representative" provisions in their national legislation, which are particularly restrictive in China where a representative must have technical competence with handling substances and have certain amount of restricted capital (Simpson 2012). Sanctions for violating rules vary as well as appeal procedures. While in some countries a Board of Appeals were established (China, Japan, Taiwan) in other violation of import rules is in jurisdiction of court administrative review action (Simpson 2012). There are also in some cases regulations which allow joint submissions and data sharing (China and Turkey). Other countries look at the ECHA SVHC list in selection of their own hazardous chemicals.

If the amount of export to the EU can only partially reveal the EU REACH on foreign companies, what are the other options? National authorities nor chemical councils or associations, they does not usually collect data on the number of companies exporting chemicals to the EU.[1] Different way how to estimate the scope is to look at the number of registrations or the only representatives established under the requirement article 8 of REACH. According to ECHA, as of 18th November 2016 there were 1818 registrations made by the only representative for 1081 substances and other 3160 registrations made by importers, covering 1658 substances (see Table 10.2).

However, this is only raw estimation as the only representative (OR) may represent several countries and enterprises at once. Moreover, It is not mandatory to include information about party representing into the registration dossier. Another way is to look at the number of only representative services providers and collect data about their clients. Here the problem is that OR market is very diverse. There are some OR service providers but the total number can be hard to identify as some daughter companies are natural OR services providers for their mother companies. The only statistics available is those from the Only Representative Organization – an industry association composed of more than 30 OR providers following high

[1] Author contacted 23 chemical associations or chemical councils with very limiter response.

Table 10.2 No. of registration made by supply chain

Supply chain role	No. of registrations	%	No. of substances
Manufacturer	1797	25	1399
Manufacturer and Importer	418	6	385
Importer	3160	44	1658
Only Representative of a non-EU manufacturer	1818	25	1081
Total	7193	100	–

Source: ECHA (2016), p. 1

standards in their practice. Unfortunately, not all OR providers are able to follow these standards and thus are not part of ORO.

According to ORO data their members represent more the 2 325 companies, from which majority of 1727 manufacturers are based in Asia (ORO 2015). These numbers are not surprising despite they represent only unknown part of the complex picture.

Another way how to reveal the impact of REACH on non-EU companies is using data from SEIF (Substance Information Exchange Fora). However, even here the picture will be not complex. The potential registrants of a certain phase-in substance might be visible but this is not the case of new substances. Moreover, some importers are represented by OR who acts as registrants and thus the non-EU company will be not visible to other SEIF participants.[2] In other words, it is impossible to exactly find number of non-EU companies affected by REACH and their place of residence. The number of registrations or the ORO customers my give us only very general idea about their origin. In other words, the impact of REACH beyond borders is hardly measurable in numbers.

10.2 Theoretical Implications

Why are some countries shifting its own chemical regulation closer to REACH and some not? The answer may lie within Europeanization theory. The question of adaptation in connection to REACH was researched by several authors.

Tachwa Lee (2015) analyzed factors leading to policy adaptation in the case of South Korea. Lee pointed out that in the case of South Korea policy convergence occurred due to both national (concerns about competitiveness and accidents) and international factors, such as diffusion and regulatory competition (Lee 2015: 16). Despite there were considerable contacts between Korean authorities and EU officials from the European Chemicals Agency and the EU Commission (See Biedenkopf 2013 cited in Lee 2015: 13), the nature of Europeanization was indirect as there was no legal obligation for Korea to adopt REACH. Lee also works with the

[2] For this point I am thankful to Dr. Thomas Blumenthal from ExperChem Ltd., Weinheim.

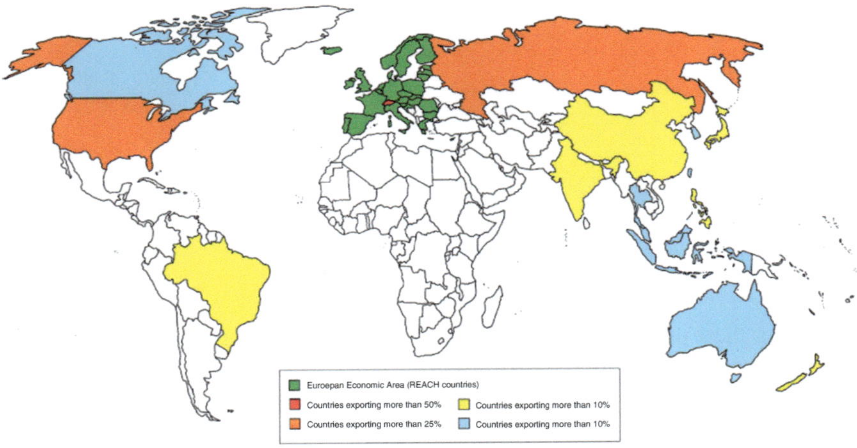

Map 10.1 Chemical export to the EU (% share)
Source: Author, based on WITS 2015 data

chemical industry and NGO support of the reform and active attitude of the government which created the supportive environment to policy reform (Lee 2015: 13).

It seems that the level of chemical industry support to reform matters as shown in the case of Japan. Japanese chemical industry did not support EU REACH-like reform and instead preferred approach similar to Canadian regulatory framework which better fits Japanese domestic regulatory practice (Naiki 2010: 191). In this sense, similarity of domestic structures should be considered in the case of adoption to REACH.

Naiki also points out, that another important factor which undermined the pressure to adopt complex chemical regulation is the possible disclosure of information in the EU (Naiki 2010: 193). Why non-EU countries should introduce complex and demanding regulation which might have negative impact on competitiveness and employment when all necessary information will be soon publicly available from the European Chemical Agency? In this sense, EU chemical industry and EU importers pays for data gathering and take the burden while benefits may be used by all actors around the World. On the other side, the "wait and see" attitude of regulators all around the World may soon be changed to direct reference to REACH annexes or at least information collected by REACH and support a technical convergence between regulatory regimes (Map 10.1).

The question is, how factors influencing REACH may be "translated" into Europeanization theory. As noted earlier in the first chapter, Europeanization may be divided to direct and indirect and later to voluntary and coercive based on the attitude of domestic actors (Bache 2002: 11). What actors shall be included? There are both attitudes: supporting stronger regulations (mainly NGOs) and weaker regulation (Industry). The position of government in the legislative proposal shall consider both attitudes, however due to high diversity between countries and their administrative practices and various levels of democracy, this perception will certainly vary among countries. From the policy outcome it is difficult to recognize which position was considered in greater extent as other socio-economic aspects and existing experience are to be considered as well. To map actual domestic environment requires deep analysis in order to determine, whether Europeanization occurs in a voluntary or coercive way and how it depends on the actors.

Nevertheless, we can assume that the chemical industry will be opposing compulsory regulations limiting the profit and increasing obligations towards the safety, health and the environment. On the other side, legislators in the non-EU countries are not bound by EU law and shared sovereignty. The EU is not actively pushing them to adopt REACH-like regulation except necessary compliance for exporting companies to meet the EU standard. In this sense, EU influence is indirect and Europeanization is taken on a voluntary way. This is very close to the "framing" approach in the Europeanization theory, despite authors intended this type to characterize relation between EU and its member states (Knill and Lehmkuhl 2002: 259).

Domestic revisions in countries open also the issue of a "misfit" which is in all countries are well visible as domestic regulatory regimes differ from REACH. The misfit logic, that higher misfit creates higher pressure for adaptation (Börzel and Risse 2000: 5) is not valid in the case of REACH which has implications for the "misfit model" of Europeanization (Beichelt 2008: 3). However, despite countries did not adopted REACH-like provisions, they in many cases recently improved existing chemical regulations by shifting the approach towards a REACH direction (partial adaptation). In this sense misfit remains but the states went closer to close the gap and in some cases covered some spaces within the misfit. EU created new international standard which set new normative.

We can observe rather a sporadic adaptation of non-EU countries (EU candidates excluding) to REACH as domestic driven process only slightly influenced by Europeanization and globalization. However, despite dominance of domestic factors in non-adaptation to REACH we can expect that globalization and Europeanization took its effects as well but to a different extent. This logic fits well into the "Enlarged Transformation Model" which involves territorial and systemic contexts within the theory of Europeanization (Beichelt 2008: 7) as showed on the picture 10.1.

This model allows to include various dimensions which have influence on the political process. All states dealt in this book made considerable changes vis-a-vis demands coming from the international level and implemented international obligations under various treaties, starting from Minamata Convention on Mercury to Stockholm Convention or GHS. The process of globalization via international conventions seems to be successful as almost all considered states implemented or are implementing conventions and GHS.

Domestic reforms aimed at increasing chemical safety and human health or environment protection shall be considered in the regional context. It is not surprising that the South-East Asian regulatory regimes have certain degree of similarity regarding chemical management.

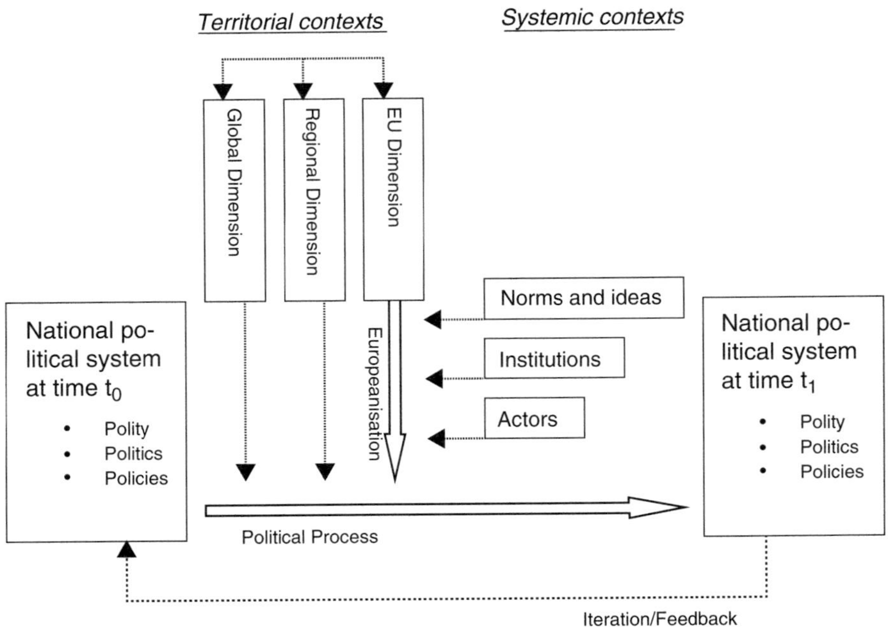

Picture 10.1 Enlarged transformation model
Source: Beichelt (2008), p. 7

The EU impact in the form of norms and ideas seems to be limited as there are both rationalist (socio-economic benefits) and constructivist arguments (logic of appropriateness) against the influence. Despite REACH being considered in the case of Russia, Japan or other states, a partially influenced implementation only occurred in Korea and Switzerland. The level of economic dependence, however, is not the only explanatory factor, as there are states with higher export importance than in the case of Korea. It seems that domestic environment in the case of Korea favored the open approach of the Korean government towards REACH like regulation (Lee 2015). In this sense economic dependence was not a satisfactory condition but one of the conditions enabling Europeanization. The importance of domestic structure and structure of economy seems to be crucial in the case of Russia and other states which have rather chosen their national way of revision.

In Sect. 1.4 there were presented eight modes of Europeanization presented by Frank Schimmelfenning (2012). The question is which mode is most suitable for Europeanization regarding REACH impact beyond borders. Europeanization may be EU driven or domestic driven. In the case of REACH EU is not pressing other governments to adopt EU-like norms nor domestic governments are rather passive in adopting EU rules (except Switzerland and South Korea). However, REACH is reality and foreign companies have to follow data gathering rules due to registration process. In this sense, Europeanization is not visible on a state level but in partial adaptation of companies and chemical industry. This is even better visible in countries such as Ukraine or Turkey where many companies adapted REACH requirements and in many other countries where REACH compliance assistance services are provided by agencies. These companies are not obliged to adopt EU safety standards and change their internal production processes, however they have to follow the REACH rule "no data, no market" and submit certain amount of information regarding the product imported to the EU.[3] At least in the terms of data gathering, Europeanization is in those cases latent, untransformed into official legal rules aimed at data gathering of those countries.

This form of Europeanization is very close to transnational externalization mode where EU is seen as a presence (Schimmelfenning 2012: 9). EU internal economic governance (such as regulation) may produce negative externalities to third countries and its societal actors. Trade interdependence creates pressures to the third countries to adopt rules in order to enter EU single market. Ignoring EU norms will create net costs (Schimmelfenning 2012: 9). There are two premises: First, the more trade with the EU, the more compliance with EU rules; and second, the benefits from adaptation must be higher than costs of ignoring the rules. This is most visible in the case of Switzerland which was unable to ignore REACH due to its export structure strongly dependent on the EU and possibly also by the structure of the market (e. g. presence of daughter companies with EU based mothers). However, also logic of appropriateness may be used as in many areas EU is seen as normative source of inspiration for dealing with domestic problems.

There are many countries with significant share of EU oriented export in chemicals but none of them fully reflected REACH on the national level. REACH has been mainly seen as an expensive alternative to domestic regulation or incompatible with existing regulatory practice and capacities. However, in the case of South Korea and Switzerland it seems that logic of consequences met the logic of appropriateness and created a unique environment where reflection of REACH led to legislative changes.

In other countries it seems that ignoring REACH presence and associated costs were considered as less costly than introducing complex REACH-like regulation and possibly undermine the performance of the domestic industry which in Asia is regionally oriented. Due to REACH complexity and its data gathering provisions third-world countries may have considered the introduction of complex regulation as unnecessary and have chosen the "wait and see" approach. However, once all substances will be registered and their properties mostly known, the environment will change in favor of common standards. The normative power of REACH together with existing experience may be translated into the real policy proposals and lead towards greater similarity within national regulatory systems and also possibly to the development of more comprehensive global chemical management system.

[3]We can not talk about compliance with REACH as REACH is fully legally binding only in European Economic Area but rather about partial adaptation to REACH requirements for non-EU importers.

Conclusion

Globalization of international trade with chemicals created new challenges which require regulation in order to protect human health and the environment. Adoption of REACH in the EU presented an important milestone in addressing challenges related to chemical management. Yet, the impact of REACH beyond EU borders is overestimated.

The influence of REACH is most visible in the EU neighborhood and depends mainly on the interaction between EU and local governments which expects early membership in the EU or some other advantages from the EU, such as access to the market. This is notably the case of Montenegro, Serbia, Turkey or Kosovo, where REACH had strong impact. Beyond EU and its potential candidates the impact of REACH is much weaker. Out of 17 analyzed countries only two made considerable changes in the light of REACH. The case of Switzerland is not surprising due to its geographies locating in the middle of Europe. Swiss authorities enacted legislation which to the high extent copy REACH provisions.

In the USA TSCA has been amended in 2016 with Lautenberg Act. It goes at least in seven aspects closer to REACH. However, as in other reform proposals, it is mainly evolutionary corrective to national legislation than new revolutionary approach. Chemical regulation in Russia reflects REACH to higher extent, especially registration of chemical follows similar attitude. Despite initial enthusiasm, deeper reflection of REACH provisions and possible inspiration to national regulation was undermined by Russian administrative and financial reality which resulted in partial improvements of existing legislation. Similar attitude is present in China, where provisions on notification (based on tonnage volumes) reflect the logic of REACH, however overall impact of REACH is marginal.

REACH maybe influenced some minor aspects within chemical regulation of Japan, where prioritization of chemicals for assessment reflects the logic of REACH. However, from all Asian countries the case of REACH influence is strongest in South Korea. This is even more surprising due to its geographic location. In the case of South Korea more factors than just economic dependence were involved behind Europeanization process. An important role was attributed to NGOs, the response of chemical industry and consultations between regulatory experts and the

O. Filipec, *REACH Beyond Borders*, DOI 10.1007/978-3-319-54154-9

Table A1 Selected legislative amendments 2008–2015

Date	Milestone
1 January 2008	Environmental Protection and Management Act of 2008 enters into force in Singapore
March 2008	Act No. 895 amends South Korean Toxic Chemicals Control Act (TCCA)
26 December 2008	Turkey adopted the Regulation on Inventory and Control of Chemicals (CICR)
January 2009	Environmentally Hazardous Substances Notification and Registration Scheme (EHSNRS) presented in Malaysia
1 February 2009	Chemicals Ordinance on Protection against Dangerous Substances and Preparations reflects REACH in Switzerland
20 May 2009	Japan introduces Chemical Substances Control Law (CSCL)
October 2010	Order No. 7 updated Measures of Environmental Management of New Chemical Substances provisions (China)
14 April 2011	Safe Chemicals Act of 2011 (S.847) introduced in the USA
5 February 2012	General Report on REACH review adopted by the European Commission
10 April 2013	Safe Chemicals Act of 2013 (S.696) introduced in the USA
22 May 2013	Chemical Safety Improvement Act (S.1009) introduced in the USA
July 2013	Hazardous Substance Act was published for public consultations (Thailand)
11 October 2013	CLASS Regulation introduced GHS in Malaysia
11 December 2014	Presidential Order amends Toxic Chemical Substances Control Act (Taiwan)
January 2015	South Korean Toxic Chemicals Control Act is divided into Korea REACH and Chemicals Control Act (CCA)
19 February 2015	Thailand Hazardous Substances Notification (B.E. 2558) published
1 July 2015	Act 59 amended Industrial Chemical Act of 1989 (Australia)

European Commission. However, some minor changes in line with reach can be seen as well in Thailand and Taiwan.

Other countries in Asia adopted a more careful approach. While all of them changed its chemical law in the last 10 years none adopted the complex registration requirements as REACH (for major changes see Table A1). However, we can see similarities among regulatory regimes in Asia. First, all regulatory regimes are based on inventory of registered or notified chemicals. Second, only selected chemicals are subject of assessment as complex assessment is considered expensive and unnecessary. Third, failure to act in accordance with legislation may result in administrative, financial or penal sanctions, although they vary significantly. Fourth, almost all countries in the region used and implemented the GHS standard. Fifth, except Malaysia, all countries acceded and to some degree implemented the Basel, Stockholm and Rotterdam conventions. In addition, all countries signed the Minamata Convention on Mercury, which is being implemented now. Those countries are fully participation on globalized trade with chemicals.

Changes in Asia are particularly important due to global trends. Chemical export from Asia is rising. While in 2000 chemical export from the East Asian and Pacific countries was worth \$75 billion USD, in 2010 it was \$230 billion USD and in 2014

East Asian and Pacific export was worth $250 billion USD (WITS 2015). The development in Asia correlates with the development in World export. While in 2000 global chemical export was worth approximately 500 billion, in 2010 it was three times higher. In all regions, including Sub-Saharan Africa chemical export has been rising. Off-shoring of chemical production to developing World implies higher responsibility of local governments to introduce new regulations reflecting international standard. Pioneering entities such as the EU or Canada may play very important role in this transformative process.

This is particularly true for countries which are experiencing progressive build up of their national chemical regulation like India, Brazil or who are attempting to modernize its chemical regulation (Ukraine, Moldova), or correct systemic capacity problems (Malaysia). REACH or CEPA may be very good example for those countries in procedural or technical sense. Some countries already look at EU SVHC list and simply copy it into their regulatory systems.

Regulation of hazardous chemicals is mainly the domain of OECD countries, however following EU REACH the "regulatory earthquake" reached also Asian countries and regulatory improvements are considered on all continents. While BRIC countries have still a long way to improve their legislation, it could be expected, that regulatory convergence between BRIC countries and OECD countries or the EU will be rising.

Nevertheless, it is too early to talk about real global regulation and as showed in this book, the impact of EU REACH is limited and Europeanization is present only in its indirect or in area of the data gathering provisions rather latent form. Nevertheless, this does not mean that there is no point. REACH sets a normative example which has created the most comprehensive regulatory regime in the World in the field of chemicals. For this reasons, other countries will closely watch the EU experience. Information revealed under REACH will create a new environment favoring further compliance and shifting regulatory regimes closer. Maybe once regulatory authorities and political leaders will recognize that impact of hazardous chemicals on human health and the environment is global issue which requires global solution in the form of global regulatory system. Political will backed with expertise may once lead to creation of global inventories of chemicals. Once global inventories will be created, then global regulatory procedures may follow, including common chemical assessment or universal restrictions. One for all system may help to avoid duplicate tests and limit unnecessary suffering of animals. Shared global system can make information gathering cheaper and easier. Moreover, the existence of common authority will decrease the cots due to economies of scale and maximalize the utility of professional personal. Global, complex and scientific approach considering human health and the environment as the supreme values within regulation of hazardous chemicals seems to be the right answer to the challenges posed by hazardous chemicals.

This is still far from reality. However, as forms of economic integration spread around the globe and states are more engaged in free trade areas, customs unions or common markets, there is increasing and from the environmental point of view urgent importance of common rules regarding regulation of chemicals.

Bibliography

ABC News (2015) Women put an average of 168 chemicals on their bodies each day, consumer group says. ABC News, 27. 4. http://abcnews.go.com/beta/Health/women-put-average-168-chemicals-bodies-day-consumer/story?id=30615324. Accessed 14 Mar 2016

Abelkop ADK, Graham JD (2015) Regulation of chemical risks: lessons for reform of the Toxic Substances Control Act from Canada and the European Union. Pace Environ Law Rev 32(1):108–224

Act on Registration, Evaluation etc. of Chemicals (K-REACH). Ministry of Environment, Republic of Korea

Act on the Evaluation of Chemical Substances and Regulation of their Manufacture (Act No. 117 of October 16, 1973)

African Union (2013) List of countries which have signed, ratified/acceded to the Bamako Convention on the ban of the Import into Africa and the Control of Transboundary Movement and Management of Hazardous Wastes within Africa. http://www.au.int/en/sites/default/files/Bamako%20Convention.pdf. Accessed 16 Sept 2016

Allen D, Smith M (1990) Western Europe's presence in the contemporary international arena. Rev of Int Stud 16(11):19–39

Amendment of the text of and annexes I, II, III, IV, VI and VIII to the 1998. Protocol on persistent organic pollutants

American Chemistry Council (2016) ACC comments to inform EPA's rulemaking on the conduct of risk evaluations under the Lautenberg Chemical Safety Act. https://www.americanchemistry.com/Comments-on-EPAs-Development-of-LCSA-Risk-Evaluation-Rule.pdf. Accessed 19 Nov 2016

An Act to Control Toxic Substances and Hazardous and Nuclear Wastes, providing penalties for violations Thereof, and for other purposes. Republic Act No. 6969

Anastasakis O (2005) The Europeanization of the Balkans. Brown J World Aff 12(1):77–87

Applegate JS (2008) Synthezing TSCA and REACH: practical principles for chemical regulation reform. Ecol Law Q 35:721–767

Australian Government (2015) National industrial chemicals notification and assessment scheme. http://www.nicnas.gov.au/regulation-and-compliance/nicnas-handbook/handbook-main-content/which-notification-category-do-i-need-for-a-new-chemical/do-i-apply-for-a- permit-or-an-assessment-certificate. Accessed 29 Nov 2015

Bache I (2002) Europeanization: a governance approach. Paper to be presented at the EUSA 8th international biennial conference, Nashville, 27–29 March 2003

Bache I (2003) Report on the commission white paper on European Parliament (2001). http://aei.pitt.edu/1722/1/Bache.pdf. Accessed 14 Oct 2016

Bahri S (2011) Hazardous and toxic material management in Indonesia. Ministry of Environment the Republic of Indonesia. http://www.oecd.org/ehs/pfc/48725707.pdf. Accessed 2 Feb 2016

O. Filipec, *REACH Beyond Borders*, DOI 10.1007/978-3-319-54154-9

Bamako Convention on the ban on the Import into Africa and the Control of Transboundary Movement and Management of Hazardous Wastes within Africa

Basel Convention (2015a) Parties to the Basel convention on the control of transboundary movements of hazardous wastes and their disposal. http://www.basel.int/Countries/StatusofRatifications/PartiesSignatories/tabid/4499/Default.aspx. Accessed 28 Sept 2016

Basel Convention (2015b) Ban amendment to the Basel convention on the control of transboundary movements of hazardous wastes and their disposal. http://archive.basel.int/ratif/ban-alpha.htm. Accessed 17 Sept 2016

Basel Convention on the control of transboundary movements of hazardous wastes and their disposal

Bauer M, Knill CH, Pitschel D (2007) Differential Europeanization in Eastern Europe: the impact of diverse EU Regulatory Governance Patterns. J Eur Integr 29(4):405–423

Beichelt T (2008) Dimensions of Europeanization. In: Bafoil F, Beichelt T (eds) Européanisation. D'Ouest en Est L'Hartmattan. Coll. Logiques Politiques, Paris, pp 1–14

Benton-Short L, Benton JR (2008) Cities and nature. Routledge, New York

Bergeson, L., Macdougall, L. S., Cana, R., Scott, L. and Navin-Jones, M. (2010) Turkey enacts REACH-Like chemical program: what stakeholders need to know. chemical regulation reporter, 18 Jan 2010, pp 1–4

Berend K (2009) Experience and global implications of REACH. Contribution presented at Helsinki Chemicals Forum, Helsinki, 28–29 May 2009

Birry AA (2014) Fighting for a toxics-free future in Indonesia. Global business briefing, June 2014. https://chemicalwatch.com/20159/fighting-for-a-toxics-free-future-in-indonesia. Accessed 22 Nov 2015

Borneman J, Fowler N (1997) Europeanization. Annu Rev Anthropol 26:487–514

Börzel T (1999) Towards convergence in Europe? Institutional adaptation to Europeanization in Germany and Spain. J Common Mark Stud 37(4):573–596

Börzel T (2002) Pace-setting, foot-dragging, and fence-sitting: member state responses to Europeanization. J Common Mark Stud 40(2):193–214

Börzel T, Risse T (2000) When Europe hits home: Europeanization and domestic change. Eur Integr Online Papers 4(15):1–24

Börzel T, Risse T (2003) Conceptualising the domestic impact of Europe. In: Feathersone K, Radaelli C (eds) The politics of Europeanization. Oxford University Press, Oxford, pp 57–82

Bowens M, Yesilkagit K (2010) The EU as lawmaker: the impact of EU directives on national regulation in the Netherlands. Public Adm 88(1):57–74

Bradford A (2012) The brussels effect. Northwest Univ Law Rev 107(1):1–68

Bulmer S, Burch M (1998) Organizing for Europe: whitehall, the British State and the European Union. Public Adm 76(1):601–628

Burch M, Gomez, R (2003) Europeanization and the English regions. Paper presented at the ESRC seminar on Europeanization, University of Sheffield, 2 May 2003

Canadian Environmental Protection Act, 1999. S.C. 1999, c. 33

CEFIC (2006) Responsible Care® Global Charter. http://www.cefic.org/Documents/ResponsibleCare/RC_GlobalCharter2006%5B1%5D.pdf. Accessed 21 Sept 2015

CEFIC (2010) European Responsible Care® Security Code. http://www.cefic.org/Documents/ResponsibleCare/Feuillet%20RC_SecurityCode_V4.pdf. Accessed 22 Sept 2015

CEFIC (2015a) The chemical industry's commitment to sustainability.http://www.cefic.org/Responsible-Care/. Accessed 24 Sept 2015

CEFIC (2015b) 10th Edition of the European responsible care awards 2015 entries closed. Applications under review. http://www.cefic.org/Responsible-Care/Awards/. Accessed 14 Oct 2015

CHemAdvisor (2015) Brazil regulatory updates – October 2015. https://www.chemadvisor.com/news/525-brazil-regulatory-updates-october-2015. Accessed 21 Nov 2015

Chemical Convention, 1990. (C170), International Labour Organization

Chemical Safety Improvement Act. S. 1009, May 22, 2013. To reauthorize and modernize the Toxic Substances Control Act, and for other purposes

Chemical Watch (2008) Switzerland faces tough choices on REACH. https://chemicalwatch.com/592/switzerland-faces-tough-choices-on-reach. Accessed 29 Nov 2015
Chemical Watch (2009) Chemicals regulation in the middle East – an overview. Global business briefing, May 2009. https://chemicalwatch.com/2221/chemicals-regulation-in-the-middle--east-an-overview. Accessed 21 Sept 2015
Chemical Watch (2011) Brazil: leading chemicals control in South America. https://chemicalwatch.com/9408/brazil-leading-chemicals-control-in-south-america. Accessed 17 Nov 2015
Chemical Watch (2013) Mexico launches chemical substances inventory. https://chemicalwatch.com/13446/mexico-launches-chemical-substances-inventory. Accessed 21 Nov 2015
Chemical Watch (2014) Legal spotlight special – Russia: a case study in responding to REACH, February 2009
ChemSafetyPro (2015a) GHS in other countries. http://www.chemsafetypro.com/Topics/GHS/GHS_in_Other_Countries.html. Accessed 1 Oct 2015
ChemSafetyPro (2015b) Top 10 differences between K-REACH and EU REACH regulation. http://www.chemsafetypro.com/Topics/Korea/Difference_between_K-REACH_and_EU_REACH.html. Accessed 21 Sept 2015
ChemSafetyPro (2015c) Top 10 differences between Taiwan TCSCA and EU REACH. http://www.chemsafetypro.com/Topics/Taiwan/Differences_between_Taiwan_TCSCA_and_EU_REACH_Regulation.html. Accessed 21 Sept 2015
ChemSafetyPro (2015d) Chemical registration under TCSCA. http://www.chemsafetypro.com/Topics/Taiwan/Taiwan_TCSCA_Registration.html. Accessed 21 Nov 2015
ChemSafetyPro (2016) Thailand REACH and National Chemical Inventory. http://www.chemsafetypro.com/Topics/Thailand/Thailand_REACH_and_National_Chemical_Inventory.html. Accessed 21 Dec 2016
CIRS (2013) Korea REACH – the Act on the registration and evaluation of chemicals. Chemical Inspection & Regulation Service. http://www.cirs-reach.com/KoreaTCCA/Korea_REACH_The_Act_on_the_Registration_and_Evaluation_of_Chemicals.html. Accessed 11 Nov 2015
CIRS (2014a). New chemical substance notification in China. Chemical Inspection & Regulation Service. http://www.cirs-reach.com/China_Chemical_Regulation/IECSC_China_REACH_China_New_Chemical_Registration.html. Accessed 21 Dec 2016
CIRS (2014b) Japan Chemical Substances Control Law (CSCL) Chemical Inspection & Regulation Service. http://www.cirs-reach.com/Japan_CSCL/New_Japan_Chemical_Substances_Control_La w_CSCL.html. Accessed 21 Dec 2016
CIRS (2014c) Latest developments of K-REACH hearings. Chemical Inspection & Regulation Service. http://www.cirs-reach.com/KoreaTCCA/Latest_developments_of_K-REACH_from_public_hearings.html. Accessed 21 Dec 2016
CIRS (2014d) Revised Toxic Chemical Substances Control Act (TCSCA) in Taiwan. Chemical Inspection and & Regulatory Service. Available at: http://www.cirs-reach.com/Taiwan/Revised_Toxic_Chemical_Substances_Control_Act_TCSCA_Taiwan_2013.html. Accessed 21 Dec 2016
CIRS (2015) Korea Toxic Chemicals Control Act. Chemical Inspection & Regulation Service. http://www.cirs-reach.com/KoreaTCCA/Korea_Toxic_Chemicals_Control_Act_TCCA.html. Accessed 21 Dec 2016
CIRS (2016) Global chemical regulations outlook. Chemical Inspection & Regulation Service. http://www.cirs-reach.com/news-and-articles/2016-global-regulations-outlook.html. Accessed 21 Dec 2016
Commission Delegated Regulation (EU) No 1062/2014 of 4 August 2014. on the work programme for the systematic examination of all existing active substances contained in biocidal products referred to in Regulation (EU) No 528/2012 of the European Parliament and of the Council
Commission Implementing Regulation (EU) 2016/9 of 5 January 2016. on joint submission of data and data-sharing in accordance with Regulation (EC) No 1907/2006 of the European Parliament and of the Council concerning the Registration, Evaluation, Authorisation And Restriction Of Chemicals (REACH)
Contiero M (2006) Toxic lobby: how the chemicals industry is trying to kill REACH. Greenpeace International, Amsterdam

Convention concerning safety in the use of chemicals at Work
Convention on Long-range transboundary air pollution
Council Directive (EC) 2002/95 of 27 January 2003. on the restriction of the use of certain hazardous substances in electrical and electronic equipment [2003] OJ L37/19
Council Directive 67/548/EEC of 27 June 1967. on the approximation of laws, regulations and administrative provisions relating to the classification, packaging and labeling of dangerous substances
Cowles M, Caporaso J, Risse T (eds) (2001) Transforming Europe: Europeanization and domestic change. Cornell University Press, London
ČT 24 (2008) Valašskomeziříčská chemička DEZA tajila únik naftalenu. Česká Televize. Aired 8 Nov 2008. http://www.ceskatelevize.cz/ct24/regiony/1434833-valasskomeziricska-chemicka-deza-tajila-unik-naftalenu. Accessed 23 Nov 2016
Cunningham WP, Cunningham MA (2015) Principles of environmental science. Additional case studies. Second edition. http://highered.mheducation.com/sites/0072919833/student_view0/chapter13/additional_case_studies.html. Accessed 11 Nov 2015
Decision BC 11/1, Follow-up to the Indonesian-Swiss country-led initiative to improve the effectiveness of the Basel Convention
Decree 591 of the State Council of China in 2011.: Regulations on the safe management of chemicals in China
Denison RA (2009a) Ten essential elements in TSCA Reform. Environ Law Report 39(1):1–9
Denison RA (2009b) Statement of Richard A. Denison, Ph.D. Senior scientist environmental defense fund before the U.S House of Representatives Committee on Energy and Commerce Subcommittee on Commerce, Trade, and Consumer Protection at a Hearing on Revisiting the Toxic Substances Control Act of 1976. Washington, 26 February 2009. http://www.edf.org/sites/default/files/9295_Denison_testimony_Toxics_Act_0.pdf. Accessed 12 Dec 2016
Denison RA (2016) New chemicals under the new TSCA: growing pains now, but a stronger system going forward. http://blogs.edf.org/health/2016/12/02/new-chemicals-under-the-new-tsca-growing-pains-now-but-a-stronger-system-going-forward/#more-5860. Accessed 23 Dec 2016
DEUT (2015) Improving chemicals management. Delegation of the European Union to Turkey. Accessed 9 Aug 2017. Available at: http://avrupa.info.tr/eu-projects-at-a-glance/business-environment/improving-chemicals-management.html
DG TRADE (2015a) European union trade in goods with Taiwan. European Commission. http://trade.ec.europa.eu/doclib/docs/2006/september/tradoc_113452.pdf. Accessed 15 May 2016
DG TRADE (2015b) European union trade in goods with Australia. European Commission http://trade.ec.europa.eu/doclib/docs/2006/september/tradoc_113346.pdf. Accessed 29 Nov 2015
Diez T, Stetter S, Albert M (2006) The European union and border conflicts: the transformative power of integration. Int Organ 60(3):563–593
Directive 98/8/EC of the European Parliament and of the council of 16 February 1998. concerning the placing of biocidal products on the market
Directive 1999/45/EC of the European Parliament and of the Council of 31 May 1999. concerning the approximation of the laws, regulations and administrative provisions of the member states relating to the classification, packaging and labelling of dangerous preparations
Directive 2000/60/EC of the European Parliament and of the Council of 23 October 2000. establishing a framework for community action in the field of water policy
Directive 2008/56/EC of the European Parliament and of the Council of 17 June 2008. establishing a framework for community action in the field of marine environmental policy
Dolowitz D, Marsh D (1996) Who learns what from whom: a review of the policy transfer literature. Pol Stud 44(2):343–357
ECHA (2011a) Building up capacities. Newsletter No. 4, August 201. European Chemicals Agency. Accessed 9 Aug 2017. Available at http://newsletter.echa.europa.eu/home/-/newsletter/entry/4_11_interview_serbia;jsessionid=9C6E2EE90AD2A5BC3B433832865ADD49.live1
ECHA (2011b) Preparing for EU membership. Newsletter No. 5, October 2011. European chemicals agency. Accessed 9 Aug 2017. Available at http://newsletter.echa.europa.eu/home/-/news-

letter/entry/5_11_preparing-for-eu;jsessionid=AAD5CE12B7766BF04F789F27DCFA7837.live2
ECHA (2012) Echa Prepares Balkan region for EU Accession. Newsletter No. 4, August 2012. European Chemicals Agency. Accessed 9 Aug 2017. Available at http://newsletter.echa.europa.eu/home/-/newsletter/entry/4_12-balkan;jsessionid=5AA4738D806BE6AE6354D64035CD6278.live2
ECHA (2014) The registration dossier. European Chemicals Agency. http://echa.europa.eu/regulations/reach/substance-registration/the-registration-dossier. Accessed 24 June 2016
ECHA (2015) Guidance for implementation of Regulation (EU) No 649/2012 concerning the export and import of hazardous chemicals. European Chemicals Agency http://echa.europa.eu/documents/10162/21784135/guidance_pic_en.pdf. Accessed 13 Mar 2016
ECHA (2016) Registration statistics deadline 2018. European Chemicals Agency. https://echa.europa.eu/documents/10162/13629/reach_2018_result_stats_en.pdf. Accessed 13 Nov 2016
ECRAN (2014) Workshop report activity 2.8.2 capacity building on compliance with chemical legislation (2nd Regional Workshop). Environmental and climate regional accession network. Tirana, 2–4 December 2014. Author's Archive
EMB (2015) Chemical management section. Environmental management bureau. http://www.emb.gov.ph/portal/chemical/Permitings/Pre-ManufacturePre-ImportationNotification.aspx. Accessed 29 Nov 2015
Enhert T (2008) The European chemicals agency–beyond the scope of Meroni doctrine? Risk regulation at international, European and national level. Master thesis, University of Maastricht
Environmental Protection and Management Act of 2008. Cap 94A, Rg 4. Environmental protection and management (hazardous substances) regulations
EPA (2015a) Persistent organic pollutants: a global issue, a global response. Environmental Protection Agency. http://www2.epa.gov/international-cooperation/persistent-organic-pollutants-global-issue-global-response. Accessed 17 Nov 2015
EPA (2015b) How to determine if a substance is hazardous. Environmental Protection Agency. http://www.epa.govt.nz/hazardous-substances/about/what-is-hs/Pages/default.aspx. Accessed 19 Nov 2015
EPA (2015c) Substances exempt from the HSNO Act. Environmental Protection Agency. http://www.epa.govt.nz/hazardous-substances/about/what-is-hs/Pages/Exempt-substances.aspx. Accessed 29 Nov 2015
EPA (2015d) Approvals and group standards for hazardous substances. Environmental Protection Agency. http://www.epa.govt.nz/hazardous-substances/approvals/Pages/default.aspx. Accessed 29 Nov 2015
European Commission (2013a) Commission staff working document. The former Yugoslav Republic of Macedonia 2013 progress report. Accompanying the document communication from the commission to the European Parliament and the council. Enlargement strategy and main challenges 2013–2014. COM(2013)700 final. Brussels, 16. 10. 2013. Accessed 9 Aug 2017. Available at: http://ec.europa.eu/enlargement/pdf/key_documents/2013/package/mk_rapport_2013.pdf
European Commission (2013b) Screening report Montenegro (2013) Chapter 27 – Environment and climate change. European Commission, 28 November 2013. Accessed 9 Aug 2017. Available at: http://ec.europa.eu/enlargement/pdf/montenegro/screening_reports/screening_report_montenegro_ch27.pdf
European Commission (2014) The former Yugoslav Republic of Macedonia progress report. DG Enlargement, October 2014. Accessed 9 Aug 2017. Available at: http://ec.europa.eu/enlargement/pdf/key_documents/2014/20141008-the-former-yugoslav-republic-of-macedonia-progress-report_en.pdf
European Parliament (2001) Report on the commission white paper on strategy for a future chemicals policy (COM(2001) 88 – C5-0258/2001 – 2001/2118(COS)) from 17 October 2001. Available at: http://www.europarl.europa.eu/sides/getDoc.do?pubRef=-//EP//TEXT+REPORT+A5-2001-0356+0+DOC+XML+V0//EN&language=ro
Fallström Mujkić P (2012) Chemicals legislation changing in Asia. ECHA Newsletter 3, June 2012

FAO (2015) International code of conduct on the distribution and use of pesticides. Food and Agriculture Organization of the United Nations. http://www.fao.org/docrep/x5562e/X5562e0a.htm. Accessed 28 Dec 2015
Featherstone K, Kazamias G (2000) Introduction: Southern Europe and the Process of "Europeanization". South Eur Soc Polit 15(2):1–24
FEON (2007) Impact of REACH on Switzerland. Switzerland's options to Act, and consequences for the environment, health and the economy. Federal office for the environment. environmental studies 33
Finkel M (2007) Malaria. National Geographic, July 2007. http://ngm.nationalgeographic.com/ngm/0707/feature1/text4.html. Accessed 17 Nov 2015
Flockhart T (2010) Europeanization or EU-ization? The transfer of European norms across time and space. J Common Mark Stud 48(4):787–810
Gaskell S, Persson M (2010) Still out of control? measuring eleven years of EU regulation. Second edition. Open Europe. http://www.openeurope.org.uk/Content/documents/Pdfs/stilloutofcontrol.pdf. Accessed 14 Oct 2016
Gawrich A, Melnykovska I, Schwickert R (2010) Neighborhood Europeanization through ENP: the case of Ukraine. J Common Mark Stud 48(5):1209–1235
Gibson RB (1999) Voluntary initiatives: the new politics of corporate greening. Broadview Press, Ontario
Globally Harmonized System of Classification and Labelling of Chemicals (GHS). United Nations Economic Council for Europe)
Goetz KH, Hix S (2001) Europeanised politics. European integration and national political systems. Frank Cass, London
Government regulation of the Republic of Indonesia No. 74/2001 on hazardous and toxic substance management
Gunter, Valerie J. and Harris, Craig K. (1998) Noisy Winter: The DDT Controversy in the Years Before Silet Spring. Rural Sociology 63(2):179–198
Hang TTN (2011) Europeanization: simply a top-down process? Marmara J Eur Stud 19(1):135–151
Haverland M (2000) National adaptation to European integration: the importance of institutional veto points. J Public Policy 20(1):83–103
Haverland M (2005) Does the EU cause domestic developments? The problem of case selection in Europeanization research. Eur Integr Online Pap 9(2):1–14
Hazardous substances Act B. E. 2535 (Bhumibiol Adulyadej Rex)
Hazardous Substances and New Organisms Act 1996
He N (2013) Taiwan updates toxic chemical substance list. Chemical News, 28 January 2013, Chemlinked. https://chemlinked.com/news/chemical-news/taiwan-updates-toxic-chemical-substance-list. Accessed 23 Oct 2015
Heyvaert V (2008) The EU chemicals policy: towards inclusive governance? LSE Law, society and economy working papers 7/2008. London school of economics
H.R.2576–Frank R. Lautenberg chemical safety for the 21st century Act
HSD (2012) Guidance for the industry on the notification and registration scheme of Environmental Hazardous Substances (EHS) in Malaysia. Second edition. Hazardous Substances Division
ILO (2003) Safety in the use of chemicals at work. International Labour Office, Geneva
ILO (2015) Ratifications of C 170–Chemicals Convention, 1990 (No. 170). http://www.ilo.org/dyn/normlex/en/f?p=1000:11300:0::NO:11300:P11300_INSTRUMENT_ID:312315. Accessed 13 Feb 2016
Index Mundi (2014) Taiwan economy profile 2014. http://www.indexmundi.com/taiwan/economy_profile.html. Accessed 26 Apr 2015
Industrial Chemicals (Notification and Assessment) Act of 1989
IPHA (2012) National profile of chemicals management in Albania. Update 2012. Institute of Public Health of Albania: Tirana, November 2012. Accessed 9 Aug 2017. Available at: http://cwm.unitar.org/national-profiles/publications/cw/np/np_pdf/Albania_National_Profile_final.pdf

Jegelevicius L (2013) Ukraine looks forvard to aligning legislation with REACH. Chemical Watch, 3. 12. 2013. Accessed 9 Aug 2017. Available at: https://chemicalwatch.com/17481/ukraine-looks-forward-to-aligning-legislation-with-reach

Jegelevicius L (2014) Moldova outlines chemicals management alignment with EU. Chemical Watch, 24. 3. 2014. Accessed 9 Aug 2017. Available at: https://chemicalwatch.com/18827/moldova-outlines-chemicals-management-alignment-with-eu

Jetro Thailand (2015) Hazardous Substances Act and regulations issued under the Hazardous Substances Act. https://www.jetro.go.jp/thailand/e_survey/hazardousact.html. Accessed 22 Nov 2015

Joint Committee (2008) Joint committee report. Ministry of Economy, Trade and Industry. http://www.meti.go.jp/policy/chemical_management/english/files/joint_ committee_report.pdf. Accessed 16 Nov 2016

King A, Lenox M (2000) Industry self-regulation without sanctions: the chemical industry's responsible care program. Acad Manag J 43(4):698–716

Kinkela D (2011) DDT and the American Century: global health, environmental politics and the pesticide that changed the world. The University of North Carolina Press, Chapel Hill

Kinnunen L (2008) Russia harmonises its chemical legislation with reach. Kemia-Kemi 35(3):38–39

Knill CH, Lehmkuhl D (2002) The national impact of European Union regulatory policy: three Europeanization mechanisms. Eur J Polit Res 41(2):255–280

Kongprachaya N (2015) Thailand's Department of Industrial Works Publishes notification concerning the list of hazardous substances requiring notification. ChemAdvisor, 25 March 2015. http://www.chemadvisor.com/news/460-thailand-s-department-of-industrial-works-publishes-notification-concerning-the-list-of-hazardous-substances-requiring-notification. Accessed 25 June 2015

Kongprachaya N (2016) Thailand's existing chemicals inventory and proposed chemicals management regulations. ChemAdvisor, 16 August 2016. https://www.chemadvisor.com/news/624-thailand-s-existingchemicals-inventory-and-proposed-chemicals-management-regulations. Accessed 23 Dec 2016

Krishnan D, Salvi R, Shriya R (2013) Environmental legislation for industry chemicals in India. Finpro, May 8, 2012

Ladrech R (1994) Europeanization of domestic politics and institutions: the case of France. J Common Mark Stud 32(1):69–88

Ladrech R (2010) Europeanization and national politics. Palgrave Macmil- lan, Basingstoke

Lavenex S, Ucarer EM (2004) The external eimension of Europeanization: the case of immigration policies. Coop Confl 39(4):417–443

Lee T (2015) Policy convergence between Korea and the European Union in the field of chemical controls. Asia Eur J 13(4):395–412

Levitsky S, Way L (2006) Linkage versus leverage: rethinking the international dimension of regime change. Comp Polit 38(4):379–400

Lipman Z (2002) A dirty dilemma: the hazardous waste trade. Harv Int Law Rev 23(4):67–71

Mākelä P (2013) Brazil interested in learning from ECHA's experiences. ECHA Newsletter 6, 2013, pp 22–23

MENR (2008) National profile for chemicals management in the Republic of Moldova. Ministry of ecology and natural resources, Chisinau 2008. Accessed 9 Aug 2017. Available at: http://www.moldovapops.md/app/includes/files/NPCM%20Moldova%20Eng.pdf

Naiki Y (2010) Assessing policy REACH: Japan's chemical policy reform in response to the EU's REACH regulation. J Environ Law 22(2):171–195

Niemiec CH (2015a) Malaysia proposes chemicals management act and inventory. Chemical watch, 14 October 2015. https://chemicalwatch.com/42941. Accessed 18 June 2016

Niemiec CH (2015b) Thailand seeks information on chemicals. Chemical watch, 5 March 2015. https://chemicalwatch.com/23062/thailand-seeks-information-on-chemicals. Accessed 27 Nov 2016

Nye J (2004) Soft power: the means to success in world politics. Public Affairs, New York

Occupational Safety and Health (Classification, Labelling and Safety Data Sheet of Hazardous Chemicals) Regulations 2013. P.U. (A) 310

OEHHA (2013) Proposition 65 in plain language. Office of Environmental Health Hazard Assessment. California Environmental Protection Agency. http://oehha.ca.gov/prop65/pdf/P65Plain.pdf. Accessed 27 Oct 2016

Olsen JP (2002) The many faces of Europeanization. J Common Mark Stud 40(5):921–952

ORO (2015) ORO facts & figures. Only representative organization. http://www.onlyrepresentative.org/about-oro/facts-figures. Accessed 22 Dec 2016

OSPAR Commission (2015) Chemicals for priority action. http://www.ospar.org/work-areas/hasec/chemicals/priority-action. Accessed 14 May 2016

Prakash A (2000) Responsible care: an assessment. Bus Soc 39(2):183–209

Proposition 65. Safe drinking water and Toxic Enforcement Act of1986. Office of Environmental Health Hazard Assessment

Protocol on Persistent Organic Pollutants (POPs). United Nations Economic Commission for Europe

Provisions on Environmental. Administration of New Chemical Substances Management in China. Ministry of Environmental Protection Order No. 7 from January 19, 2010

Purwandari YK (2014) Chemicals/hazardous substances management in Indonesia. Ministry of environment, republic of Indonesia. http://www.chemical-net.info/pdf/20140220_Seminar2_eng.pdf. Accessed 22 Nov 2015

Purwandari, YK (2015) Hazardous substances management in Indonesia. Ministry of environment and forestry, republic of Indonesia. Available at: http://chemicalnet.env.go.jp/pdf/20150220_Seminar2_eng.pdf (17 Mar 2016)

Radaelli CM (2000) Whither Europeanization? Concept stretching and substantive change. Eur Integr Online Pap 4(8):1–28

Radaelli CM (2004) Europeanisation: solution or problem? Eur Integr Online Pap 8(16):1–26

REACH Centre (2015) New Turkish chemicals regulation [online]. REACH Centre [cit. 2015-7-22]. Available from (22 July 2015)

Regulation (EC) No 1013/2006 of the European Parliament and of the Council of 14 June 2006a. on shipments of waste

Regulation (EC) No 1272/2008 of the European Parliament and of the Council of 16 December 2008. on classification, labelling and packaging of substances and mixtures, amending and repealing Directives 67/548/EEC and 1999/45/EC, and amending Regulation (EC) No 1907/2006

Regulation (EC) No 1488/94 as well as Council Directive 76/769/EEC and Commission Directives 91/155/EEC, 93/67/EEC, 93/105/EC and 2000/21/EC

Regulation (EC) No 1907/2006 of the European Parliament and of the Council of 18 December 2006b. concerning the Registration, Evaluation, Authorisation and Restriction of Chemicals (REACH) establishing a Eu- ropean Chemicals Agency, amending Directive 1999/45/EC and repealing Council Regulation (EEC) No 793/93 and Commission Regulation (EC) No 1488/94 as well as Council Directive 76/769/EEC and Com- mission Directives 91/155/EEC, 93/67/EEC, 93/105/EC and 2000/21/EC

Regulation (EU) No 528/2012 of the European Parliament and of the Council of 22 May 2012a. concerning the making available on the market and use of biocidal products

Regulation (EU) No 649/2012 of the European Parliament and of the Council of 4 July 2012b. concerning the export and import of hazardous chemicals

Reihlen A, Ruut J, Engewald P, Fammler H, Moukhametshina E (2010) The Russian system of chemical management. Current Understanding. Baltic Environmental Forum Group, Hamburg

Report on the Commission White Paper. on Strategy for Future Chemicals Policy (COM(2001)88-C5–0258/2001–2001/2118(COS)) from 17. October 2001

Report on the operation of Directive 67/548/EEC on the approximation of the laws, regulations and administrative provisions relating to the classification, packaging and labelling of dangerous substances; directive 88/379/EEC on the approximation of the laws regulation and administrative provisions relating to the classification, packaging and labelling of dangerous

preparations; regulation (EEC) 793/93 on the evaluation and control of the risks of existing substances; directive 76/769/EEC on the approximation of the laws, regulations and administrative provisions on the marketing and use of certain dangerous substances and preparations. 1998. final. Brussels: European Commission
Roberts G (2012a) Indian chemicals ministry calls for “REACH-like legislation”. Chemical watch, 11 April 2012. https://chemicalwatch.com/10739/indian-chemicals-ministry-calls-for-reach-like-legislation. Accessed 29 Nov 2015
Roberts G (2012b) Malaysian environment ministry drops registration obligation. Chemical watch, 29 October 2012. https://chemicalwatch.com/12769/malaysian-environment-ministry-drops-registration-obligation. Accessed 29 Nov 2015
Rotterdam Convention (2015a) History of the negotiations of the Rotterdam convention. http://www.pic.int/TheConvention/Overview/History/Overview/tabid/1360/language/en-US/Default.aspx. Accessed 22 Oct 2016
Rotterdam Convention (2015b) Frequently asked questions about the Rotterdam convention. http://www.pic.int/TheConvention/Overview/FrequentlyAskedQuestions/tabid/1062/language/en-US/Default.aspx. Accessed 22 Oct 2016
Safe Chemicals act of 2011. (S. 847)
Safe Drinking Water and Toxic Enforcement Act of 1986. Health and Safety Code, Chapter 6,6 Sections 25249.5 trough 25294.13
Santilo D, Johnston P (2006) Fatal flaws. Effect thresholds and “adequate control” of risks: the fatal flaws in the Council position on Authorization within REACH. Greenpeace Report 2006. Greenpeace International
Schierow L-J (2013) The Toxic Substances Control Act (TSCA): a summary of the act and its major requirements. Congressional Research Service, 1 April 2013. http://fas.org/sgp/crs/misc/RL31905.pdf. Accessed 17 May 2016
Schimmelfenning F (2012) Europeanization beyond Europe. Living Rev Eur Govern7(1):1–31
Schimmelfennig F, Sedelmeier U (2005) Introduction: conceptualizing the Europeanization of Central and Eastern Europe. In: Schimmelfenning F, Sedelmeier U (eds) The Europeanization of Central and Eastern Europe. Studies in Political Economy. Cornell University Press, Ithaca, pp 1–28
Schörling I (2004) REACH – the only planet guide to the secrets of chemicals policy in the EU. What happened and Why? The Greens, Brussels
SCHP ČR (2014) 20 years: responsible care in the Czech Republic. Svazchemického průmyslu ČR. http://www.schp.cz/images/Soubory/Responsible_Care/140926_20let_SCHP_300.pdf. Accessed 1 Jan 2015
Selin H (2007) Coalition politics and chemical management in a regulatory ambitious Europe. Glob Environ Polit 7(3):63–93
Shapiro S (2013) Toxic chemical reform must help, not hinder, states and victim’s rights. Huffington Post, 30 October 2013. http://www.huffingtonpost.com/sidney-shapiro/toxic-chemical-reform-mus_b_4170038.html. Accessed 6 Oct 2014
Sharma V, Ashutosh K, Dhawan A (2012) Nanomaterials: exposure, effects and toxicity assessment. The National Academy of Sciences, India. http://link.springer.com/article/10.1007%2Fs40011-012-0072-7#page-1. Accessed 1 Oct 2014
Simpson C (2012) Global developments in regulation of chemicals. Belgian Association of Chemical Distributors. Weekend de la Chimie, Liege, 14 September 2012
Stadler L (1993) Corrosion proof fittings v. EPA: asbestos in the fifth circuit: a battle of unreasonableness. Tulane Environ Law J 6(2):423–438.
State of California Department of Justice (2016) Frequently Asked Questions. Available at: https://oag.ca.gov/prop65/faq (12.3.2016)
Stockholm Convention (2015a) Status of ratifications. http://chm.pops.int/Countries/StatusofRatifications/PartiesandSignatories/tabid/252/Default.aspx. Accessed 20 Apr 2016
Stockholm Convention (2015b) The new POPs under the Stockholm Convention. http://chm.pops.int/TheConvention/ThePOPs/TheNewPOPs/tabid/2511/Default.aspx. Accessed 21 Apr 2016

Swiss Ordinance on Protection against Dangerous Substances and Prepa-rations (Chemicals Ordinance, ChemO) of 18 May 2005. Definitive

Synergies (2015) History of the synergies process. http://synergies.pops.int/Decisionmaking/Overview/SynergiesProcess/tabid/2615/language/en-US/Default.aspx. Accessed 30 May 2016

Tahiri, E. (2015) Personal correspondence with Enver Tahiri, head of chemical and waste division, ministry of environment and spatial planning, Kosovo, from 6 Aug 2015. Author's Archive

The Rotterdam Convention on the Prior Informed Consent Procedure for Certain Hazardous Chemicals and Pesticides in International Trade

Thorton JW, McCally M, Houlihan J (2002) Biomonitoring of industrial pollutants: health and policy implications of the chemical body burden. Public Health Rep 117(4):315–323

Topalovic P (2007) The Challenges in implementing responsible care. School of engineering practice. http://wbooth.mcmaster.ca/epp/publications/student/Responsible_Care.pdf. Accessed 11 Feb 2015

Toxic Chemicals Control Act. Ministry of Government Legislation

Toxic Chemical Substances Control Act, promulgated by Presidential Order on December 11, 2013. Laws & Regulations Database of The Republic of China

Toxic Substances Control Act as amended Trough P. L. 107–377, December 31, 2002

Trading Economics (2015) Taiwan exports 1981–2015. http://www.tradingeconomics.com/taiwan/exports. Accessed 12 July 2016

UNECE (2011) Globally harmonized system of classification and labelling of chemicals (GHS). Fourth Revised Edition. United Nations Economic Commission for Europe. United Nations, New York

UNECE (2015) Status of ratification of The 1979 Geneva Convention on Long-range Transboundary Air Pollution. United Nations Economic Commission for Europe. http://www.unece.org/env/lrtap/status/lrtap_st.html. Accessed 8 Jan 2016

UNEP (1989) London guidelines for the exchange of information on chemicals in international trade. United Nations Environment Programme. http://www2.unitar.org/cwm/publications/cbl/synergy/pdf/cat3/london_guidelines.pdf. Accessed 17 Feb 2016

UNEP (2010) Planet's leading chemicals and waste treaties launch reform of global environmental governance. United Nations Environmental Programme. http://excops.unep.ch/press/advisories/Bali-25Feb2010.pdf.Accessed 17 Feb 2016

Uyesato D, Weiss M, Stepanyan J, Park DY, Yuki K, Ferris T, Bergkamp L (2013) REACH's Impact on the rest of the world. In: Bergkamp L (ed) European Union REACH regulation for chemicals. Oxford University Press, Oxford

Waigani Convention to Ban the Importation into Forum Island Countries of Hazardous and Radioactive Wastes and to Control the Transboundary Movement and Management of Hazardous Wastes within the South Pacific Region

Wallersted K, Vassileva E (2013) Russia REACH set to move forward tomorrow. http://3ecompany.com/blog/?p=239. Accessed 17 Mar 2016

Wang R, Ghazali H (2010) Environmental Hazardous Substances (EHS) Notification and Registration Scheme. Department of Environment, Malaysisa. Case Study Workshop on the Chemicals Sector – From Principles to Practice. Asia-Pacific Economic Cooperation, Hiroshima, Japan, 1–2 March 2010

Warhurst M (2005) REACH, a new approach to chemicals regulation in Europe: a brief history, key features, and expected outcomes. J Eur Environ Plan Law 3(1):164–172

Wilson MP, Schwarzman MR (2009) Toward a new US chemicals policy: rebuilding the foundation to advance new science, green chemistry and, environmental health. Environ Health Perspect 117(8):1202–1209

Wirth DA (2007) The EU's new impact on american environmental regulation. Fletcher Forum World Affair 31(2):99–109

WITS (2015) World integrated trade solutions. Available at: http://wits.worldbank.org (14 Feb 2016)

Wordsworth A (2007) Chemicals policy in Canada, the European Union and the United States. European and Canadian Environmental Law: Best Practices and Opportunities for Co-operation. Canadian Environmental Law Association

Workplace Safety and Health Guidelines. Management of Hazardous Chemicals Programme. Workplace Safety and Health Council, 2011

World Bank (2013) World integrated data solutions. Public database. World Bank. Accessed 9 Aug 2017. Available at http://wits.worldbank.org

World Bank (2014) Evaluation of the EU-Turkey Customs Union. Report No. 85830-TR, March 28, 2014. Accessed 9 Aug 2017. Available at http://documents.worldbank.org/curated/en/298151468308967367/pdf/858300ESW0P1440disclosed090260140TR.pdf

Yen JH (2013) Proposed Reform of the Toxic Substances Control Act (TSCA) in the 113th Congress: S. 1009 Compared with S. 696 and Current Law. Congressional Research Service, October 23, 2013